Charles Smart

Handbook for the Hospital Corps of the U. S. Army

And State Military Forces

Charles Smart

Handbook for the Hospital Corps of the U. S. Army
And State Military Forces

ISBN/EAN: 9783337161279

Printed in Europe, USA, Canada, Australia, Japan

Cover: Foto ©berggeist007 / pixelio.de

More available books at **www.hansebooks.com**

HANDBOOK

FOR THE

HOSPITAL CORPS OF THE U. S. ARMY

AND

STATE MILITARY FORCES

BY

CHARLES SMART

Deputy Surgeon-General, U. S. A.

APPROVED BY THE SURGEON-GENERAL OF THE ARMY

NEW YORK
WILLIAM WOOD AND COMPANY
1898

COPYRIGHT, 1898
WILLIAM WOOD AND COMPANY

Surgeon General George M. Sternberg, U. S. Army.

GENERAL:—When Congress authorized the organization of a Hospital Corps for the Army, I wrote and published a Handbook for the Corps with the view of assisting its members in raising their qualifications to the higher standard required of them than of the detailed men previously on duty with the Medical Department. That the book answered its purpose is shown by the Examination papers filed by the candidates for the positious of Steward and Acting Hospital Steward.

I have the honor to submit this revised edition in the hope that it may prove of equal benefit to the men who enlist for service under present war conditions.

Remaining with the highest respect,
Your obedient servant,
CHARLES SMART,
Deputy Surgeon General.

WASHINGTON, D. C., April 30, 1898.

TABLE OF CONTENTS.

[The paragraphic index at the end should be consulted when a special subject is under consideration or inquiry.]

PART I.

HOSPITALS AND HOSPITAL DUTIES.

PAGE

CHAPTER I.—THE POST HOSPITAL AND THE HOSPITAL CORPS. Routine duties; reports and papers; etc., . . 1

CHAPTER II.—ACTIVE SERVICE IN THE FIELD. Consolidation for field service; division field hospitals; ambulance companies and first aid stations. Service on the march and on the battle-field; during retreat; reorganization after disasters. Articles of the Geneva Convention. Reports and papers, ... 16

CHAPTER III.—SANITARY CARE OF CAMPS. Selection of sites; prevention of camp diseases; quarters; general police and sanitary government, ... 39

CHAPTER IV.—GENERAL HOSPITAL SERVICE. Base and general hospitals; their construction, water-supply, sewerage, personnel, administration, etc., ... 70

PART II.

ANATOMY AND PHYSIOLOGY.

CHAPTER I.—THE LOCOMOTOR SYSTEM. The bones, joints, and muscles, ... 89

CHAPTER II.—THE SYSTEM OF ORGANIC LIFE. The blood

and its circulation; digestion and absorption; excretion by the lungs, skin, and kidneys; animal heat, . . . 105

CHAPTER III.—THE ADMINISTRATIVE SYSTEM. The brain, spinal cord, nerves, ganglia, and organs of the senses, . 144

PART III.

THE SPECIAL DUTIES OF THE HOSPITAL CORPS.

CHAPTER I.—MANAGEMENT OF ACCIDENTS. First aid and transportation to hospital; preparations for surgical operations; anæsthetics; cleanliness; diet; temperature; bedsores; characteristics of urine, etc., ... 155

CHAPTER II.—SHOCK, REACTION, AND INFLAMMATION. Shock, reaction, inflammatory congestion, and symptomatic fever; treatment by rest, position, cold, heat, warmwater dressings, poultices, leeching, cupping, counter-irritation, and general measures. Chronic inflammations; treatment, ... 176

CHAPTER III.—SPECIAL INFLAMMATIONS. Burns; scalds of the throat; frostbite; chilblains; contusions; sprains; abscesses; gumboils and minor surgery of the teeth; boils; carbuncles; whitlows; corns; bunions; blisters on the feet, and chafings, ... 188

CHAPTER IV.—WOUNDS. Incised; lacerated; contused; gunshot. First aid in shock and bleeding, and the field dressing of gunshot injuries. Infected wounds; erysipelas and gangrene; arrow wounds; dissection wounds, etc. Glanders, ... 203

CHAPTER V.—HEMORRHAGE. Capillary, venous, and arterial; means of arresting, ... 219

CHAPTER VI.—WOUNDS OF THE HEAD, NECK, AND TRUNK, ... 227

TABLE OF CONTENTS. vii

PAGE

CHAPTER VII.—CONDITIONS CAUSING INSENSIBILITY. Concussion, compression, and congestion of the brain; apoplexy; sunstroke; epileptic stupor; convulsions; alcoholic stupor; opium narcotism; insensibility from cold, from asphyxia, and from heat exhaustion, . . . 232

CHAPTER VIII.—ARTIFICIAL RESPIRATION, and the cases in which it should be used, ... 241

CHAPTER IX.—FOREIGN BODIES. In the eye, nose, ear, trachea, pharynx, stomach, etc., ... 248

CHAPTER X.—FRACTURES, their symptoms and treatment; of the skull, lower jaw, spine, ribs, collar-bone, shoulder-blade, arm, forearm, hand and fingers, pelvis, thigh, and leg, ... 253

CHAPTER XI.—DISLOCATIONS, their symptoms and treatment; of the lower jaw, collar-bone, shoulder-joint, elbow, wrist, fingers, hip, kneecap, knee-joint, ankle, and foot, 272

CHAPTER XII.—MANAGEMENT OF CASES OF POISONING, 286

CHAPTER XIII.—DISINFECTANTS AND THE MANAGEMENT OF INFECTIOUS DISEASES. Itch, small-pox, cow-pox, chicken-pox, scarlet fever, rose rash, measles, German measles, typhoid fever, dysentery, cholera, and yellow fever, . . 300

CHAPTER XIV.—PHARMACY. Outlines of course of instruction, ... 312

CHAPTER XV.—ELEMENTS OF COOKERY, ... 313

ANALYTICAL INDEX TO PARAGRAPH NUMBERS, . . . 319

PART I.

HOSPITALS AND HOSPITAL DUTIES.

1. A hospital is the shelter or quarters provided for the sick and wounded of a command; but in an enlarged sense it includes the provision made for the cure of the disabled, with no special reference to the shelter or building that may be used. When a hospital is attached to a stationary command it is a *post* hospital; if it accompany the command on an expedition or campaign, it is a *field* hospital; if it be detached from the command, and particularly if it receive the sick and wounded of any command, it is a *general* hospital.

CHAPTER I.

THE POST HOSPITAL, AND THE HOSPITAL CORPS.

2. The regulation post hospital building at permanent military posts is of brick, arranged for 12, 24, or 36 beds, heated by hot water, and ventilated through brick flues and galvanized iron ducts; but as a matter of fact the building may be any kind of a shelter extemporized or utilized for the care of the sick and wounded.

3. The service of the post hospital is performed by members of the Hospital Corps enlisted for, and permanently attached to, the Medical Department. Enlisted men who have served one year in the line may be transferred to the Hospital Corps as privates. Married men are not accepted as recruits, nor transferred from the line for service in the corps. Candidates for enlistment should apply to a post medical officer or to a recruiting officer. Applicants who have graduated in pharmacy, who have been licensed by State boards of pharmacy, or who have had training as nurses in civil hospitals should present certificates of their special qualifications. Slight physical defects which, under existing orders, would disqualify for the line, do not disqualify for enlistment in the corps, provided they are not of such a character as would interfere with the full performance of the duties of a sanitary soldier in garrison or in the field.

4. If the candidate is accepted he is forwarded to a company or detachment of the corps for instruction in military discipline, nursing, first aid, drill, cooking, pharmacy, clerical work, field work, and the care and management of animals. The order of exercises of a company of instruction requires the early morning hours of every day except Sunday to be occupied with policing or cleaning up the hospital and the surrounding grounds, after which on every day except Saturday and Sunday half an hour is devoted to calisthenics and from three quarters of an hour to an hour each to study and litter drill; in addition to this the junior section has an hour on pharmacy on each of four days of the week, with elementary cooking on the fifth day, while the senior section at the same hour has anatomy on three days and first aid on two days. In the afternoon an hour each is given to study and to recitations on nursing and on four days to bandaging, with disinfection and the care of instruments on the fifth day.

5. When the education of the recruit is considered complete he is assigned to duty at some post, where his services are utilized as nurse, cook, or attendant, according to his special qualifications.

6. Privates who have served one year in the Hospital Corps, and graduates in pharmacy who have served six

months and have shown particular fitness, may be recommended for promotion by the surgeon. From those thus recommended acting stewards are detailed after passing examination as to physical condition, moral character, and general aptitude, and in the principles of arithmetic, in orthography and penmanship, the regulations affecting enlisted men, care of sick, ward management, minor surgery, hospital corps drill and first aid, the ordinary modes of cooking, and elementary hygiene.

7. No person is appointed a hospital steward until he has served a year as an acting steward, nor until he has shown by examination a more extensive and detailed knowledge of the above subjects than is required of the acting steward. A re-examination before his first re-enlistment may not be required if the surgeon certifies that the candidate has performed his duties efficiently; but a re-examination is called for before a second re-enlistment, after which no further examination is ordinarily required.

8. Every post is authorized by law to have a hospital steward; two, if the garrison contains six companies, and one additional for every additional six companies; there is also provided for every post of two companies, or of a single company of cavalry, an acting hospital steward, with privates at the rate of three for each post of one company, four for posts of two companies, and one man additional for every additional two companies.

9. The duties of stewards and acting stewards are, under the direction of the surgeon, to look after and distribute hospital stores and supplies; to care for hospital property; to compound and administer medicines; to supervise the preparation and serving of food; to maintain discipline in the hospital and watch over its general police; to prepare the hospital reports and returns; to supervise the duties of

the hospital corps in hospital and in the field; and to perform such other duties connected with their positions as may, by proper authority, be required of them.

10. The steward must be an efficient disciplinarian, expert clerk, accurate arithmetician, and a trustworthy pharmacist, with as much knowledge of materia medica, therapeutics, and minor surgery as will enable him to give sound advice and suitable treatment in the minor ailments and accidents which in civil life rely on the resources of domestic medicine or on the knowledge of the nearest pharmacist; in addition, he must have that higher knowledge, for use in the wards, which enables the experienced nurse to appreciate the condition of those who are seriously ill, that their improvement may be fostered and all harmful influences excluded. At small posts, during the temporary absence of the surgeon, the unforeseen casualties and even many of the exigencies of military life impose duties upon him the satisfactory performance of which may be of the first importance to the individuals concerned.

11. The daily routine of the service of a post hospital begins at reveille, when, after roll call, the wards are tidied up and breakfast is served and cleared away before *surgeon's call* is sounded. Promptly on this call the First Sergeant of each company brings his sick to hospital for inspection. The surgeon examines each man, indicating in the sergeant's book those who are to be treated in hospital and those who are to be excused from duty or portions thereof as sick in quarters, etc. Morning reports are then sent to the Adjutant's office for the information of the commanding officer. Prescriptions for those in quarters are now filled; and the *Register of Sick and Wounded* is brought up to date by the careful entry of the morning's changes. The wards are then visited and the prescription and diet orders recorded.

After this the kitchen, dining-room, and other parts of the hospital are inspected, and the regulation visit is at end. Emergency calls bring the medical officer to the hospital at any hour, and generally, when serious cases are on hand, he may be expected before retreat or tattoo. After the morning visit he attends to his patients, in the families of officers, married soldiers, laundresses, and other attachés of the garrison, and his prescriptions reach the dispensary from time to time during the forenoon. By the time these are filled the steward has posted the records, supplied the wards with needful articles of bedding, etc., given directions for the diet of the day, and provided the required supplies from his subsistence and hospital stores and hospital-fund purchases. The afternoon may be devoted to drills, exercise, or amusement, in the absence of special calls for its occupation otherwise, and the evening to study, or, at certain periods, to the preparation of official reports and papers.

12. The studies of the members of the corps are naturally such as will fit them to act intelligently in all matters relating to the management of the hospital and the sick and wounded. Every surgeon supervises the instruction of his men and the higher education of his stewards; the latter guide and direct the acting stewards, and these perform similar offices to those who serve under them. The medical officer is required by regulations to devote at least eight hours in each month to instructing the men of the corps in the duties of litter-bearers and the methods of first aid. These studies will eventually lead every capable member of the corps to a stewardship; but besides this personal influence they serve a higher end by preparing the corps for a sudden expansion in time of war. When every acting steward is qualified to undertake the duties of steward, and every private ready to step into a higher position, the expansion

of the command can be effected by merely recruiting for the lowest grade.

13. The surgeon is responsible for the timely and accurate rendition of the reports and papers required in the service of a post hospital; but the work, except in the case of special and professional reports, is usually performed by a member of the corps, to whom the clerical work has been assigned. For all routine reports blank forms are provided by the War Department, and full instructions are printed on each of these to insure accuracy, the said instructions having the force of Army Regulations. The reports and papers are as follows:

14. Reports and papers relating to members of the Medical Department and Hospital Corps and detached men in hospital:

A *Morning Report of the Hospital Corps*, for the information of the post commander.

A *Monthly Report*, in duplicate, Form 32, *of the Personnel and Means of Transportation of the Hospital Corps*, giving a nominal list of the members and the matrons on duty, with the changes that have taken place in their status; and a numerical list of ambulances, harness, litters, etc., one copy to the Chief Surgeon, the other to the Surgeon General direct. Changes in the status of any member of the corps, or matron, as by death, discharge, re-enlistment, or transfer to meet emergencies, are to be reported immediately, by information slip, to the Surgeon General through the office of the Chief Surgeon.

A *Monthly Personal Report* from each medical officer to the Surgeon General and Chief Surgeon on information slips, and when at stations where no post return is made to the Adjutant General by letter, giving post-office address, duty on which engaged, and the source, number, and date of orders under which acting. Similar reports are required when an officer arrives at or leaves a station. A hospital steward on furlough reports to the Surgeon General and the surgeon of his station.

Ration Returns call for the articles of subsistence for each person who has to be provided for by the medical department. Issues are made by the subsistence officer for a few days, generally ten, at a time; and at each issue settlements are effected between the companies and the hospital to prevent injustice from the sending of men to quarters after they have been drawn for on hospital returns and *vice versa*. The money value of articles on the return not drawn for use is added to the hospital fund. Money thus accruing is expended exclusively for the benefit of the sick and the members of the corps in the purchase of such articles of diet, comfort, and convenience as may be required. A *Statement of the Hospital Fund*, Form 35, is forwarded monthly to the Chief Surgeon for transmission to the Surgeon General. Provision is made on this form for a *Return of Durable Property* purchased with the fund.

Muster and Pay Rolls.[1] On the last day of February, April, June, August, October, and December muster rolls are made out in duplicate, one copy for the Adjutant General, the other to be retained; and on the last day of every month three muster and pay rolls are prepared, two for the paymaster and one to be retained. These rolls bear the names of the members and attachés of the hospital corps; soldiers in hospital, detached from their companies, are mustered on separate rolls, one set for the men of each regiment. The accounts of each soldier are settled up from time to time, and when required their status is entered on his *Descriptive List and Pay and Clothing Account*.[1] When

---

[1] Blanks from the Adjutant General's Office.

a soldier is transferred from one command to another his descriptive list must be transmitted to the officer with whose command he is next to be mustered, and all the data in the said list must be noted in the *Descriptive and Deposit Book*.[1] Clothing is drawn on *Special Requisitions* from the quartermaster and is issued by the surgeon on duplicate *Receipt Rolls*, in which is entered the money value of the articles. The commutation value of the soldier's clothing allowance constitutes a stated sum, against which the value of the issued articles is entered in a *Clothing Account Book*,[1] which is balanced June 30th and December 31st. Balances due the soldier are continued as a credit in the clothing account book, but any indebtedness to the Government is charged on the muster rolls and deducted from the pay. In the case of transfer, desertion, discharge, or death the balance due the soldier or the Government is stated on the muster rolls, descriptive list, or final statements, as may be required for the continuation or closing up of the account. Clothing burned by order, to prevent contagion, is replaced on approval of the Surgeon General. When a soldier is discharged from service he is furnished with *Final Statements*[1] in duplicate and a *Discharge*.[1] The former give an exhibit of his accounts and are the vouchers on which the paymaster settles them. When a soldier in hospital detached from his command is discharged the surgeon sends a copy of the man's descriptive list to his company commander, to enable the latter to enter on the muster roll which reports the discharge all the information needful to an understanding of the accounts. When a member of the hospital corps dies *Final Statements* are prepared as in the case of discharge, with an *Inventory of Effects*,[1] in triplicate. These, except one copy of the latter, which is retained, are forwarded

[1] Blanks from the Adjutant General's Office.

to the Adjutant General. When the effects are claimed by relatives receipts are taken in triplicate to offset the copies of the inventory; when sold by a *Council of Administration* the money is deposited with a paymaster who gives the necessary receipts. When a detached man dies his company commander is notified as in the case of discharge.

15. Reports relating to sickness and other casualties or changes in the garrison:

A *Morning Report of Sick and Wounded* for the information of the commander.

A *Monthly Report of Sick and Wounded*, Form 25, to the Chief Surgeon and the Surgeon General, which is practically a transcript from the *Register of Patients*. This blank form has spaces for *Remarks* on prevailing diseases, their causation, and the measures adopted for their prevention; but histories of cases possessing a professional interest should be transmitted in the form of *Special Reports* to the Surgeon General. When a sick or wounded soldier is transferred to another hospital a *Transfer Slip* giving full particulars of the case is forwarded with the patient. The surgeon of the receiving hospital enters the data into his register and countersigning the slip forwards it to the Surgeon General. On the occurrence of cholera, yellow fever, smallpox, or other infection liable to become epidemic a report is sent to the Chief Surgeon and Surgeon General; a *Monthly List of Patients Suffering from Epidemic Disease*, Form 27, is furnished in duplicate during the continuance of the epidemic and a detailed history of the outbreak is called for at its close. Local boards of health interested should be notified at the beginning of the epidemic.

When a soldier completes his enlistment in the Regular Army by taking the oath an *Identification Card*[1] showing the

[1] Blanks from the Adjutant General's Office.

situation and character of permanent marks and scars on his person, is forwarded direct to the Surgeon General. A card is forwarded also in the case of men received as recruits from recruiting stations and rendezvous where there is no Medical Examiner. A *Monthly Report* of the *Physical Examination of Recruits*, Form 30, is required by the Surgeon General.

16. Reports relating to the post or post hospital:

A *Monthly Sanitary Report*, Form 41, giving expression to the sanitary condition of the quarters, including all buildings belonging to the post, their drainage and sewerage systems, the character and cooking of the rations, the quantity and quality of the water supply, and the clothing and habits of the men, with such recommendations as are considered needful. This report is acted on by the post commander, who returns it to the surgeon that the action taken may be entered in the *Medical History of the Post*. The report is then forwarded through military channels to the Surgeon General.

Estimates for Repairs, Alterations, or Additions to the Post Hospital, accompanied by such drawings as are needful, are forwarded to the Surgeon General by March 1st annually. They are required to show in detail the kind and cost of the materials and labor to be procured and to what extent the labor can be performed by the troops. If no repairs are required a communication to that effect should be forwarded at the proper time. *Estimates*, distinct from those for the hospital, are made at the same time *for the Construction or Repair of Hospital Stewards' Quarters*.

A *Meteorological Report*, Form 29, is called for monthly from certain posts to be transmitted to the Surgeon General through the State office of the Weather Bureau.

17. Papers relating to medical and other supplies:

Requisitions for Medical and Hospital Supplies, Form 15, for the year beginning January 1, are forwarded in triplicate to the Chief Surgeon, who transmits one copy to the Surgeon General and another to the Supply Depot. They are made for articles of the regular supply that are or probably will be deficient, and they must show the quantity of every article on hand, whether more is wanted or not. Unexpected deficiencies are provided for on *Special Requisitions*, Form 16, in triplicate, giving a list of the articles and the quantity on hand. These are forwarded to the Chief Surgeon, who retains one copy and sends the others to the Surgeon General. In emergencies Chief Surgeons are empowered to act on special requisitions, sending one copy with their action to the Supply Depot, one to the Surgeon General with an explanation of the circumstances, and retaining one; but *Requisitions for Articles not in the Supply Table* must in all cases be forwarded to the Surgeon General. When the supplies are received the *Invoice*, Form 18, and the *Packer's List*, Form 17, are verified, and *Receipts*, Form 19, are prepared in duplicate, one copy for the issuing officer and one for the Surgeon General.

Requisitions, on the Surgeon General's Office, *for Blank Forms* for the reports and returns, etc., mentioned in this list of official papers, with the exception of those noted as furnished by other bureaus of the War Department, are forwarded when necessary.

Returns of Medical Property, Form 20, are made out in duplicate on December 31, annually, after an account of stock has been taken for the preparation of the regular requisitions, or when an officer is relieved from the duty to which the returns relate. The original is sent to the Surgeon General; the duplicate with its vouchers is retained.

They show everything received, expended, etc., and remaining on hand. Names of articles that may be expended are printed on the blank form in Roman type, of non-expendable articles in italic. Articles of the latter class, worn out or unfit for use, are condemned on *Inventory and Inspection Reports of Unserviceable Property*,[1] and these constitute the vouchers relieving from responsibility. Articles destroyed to prevent contagion are covered by the certificate of the officer responsible; articles lost or destroyed by certificate in like manner or by the certificate of an officer or the affidavit of an enlisted man or citizen personally cognizant of the circumstances.

Quarterly Returns of Quartermaster's Supplies,[2] embodying the responsibility of the surgeon for clothing drawn for issue to the men of the Hospital Corps and soldiers detached from their commands, and for ambulances, litters, tents, lamps, etc., obtained by requisition on the Quartermaster's department. They are made out in duplicate, one copy for the Quartermaster General, the other to be retained, and are accompanied with invoices for articles received, receipts for those transferred or issued, and other vouchers specially called for in cases of loss, damage, or unserviceability.

18. The *Books of Record* required to be kept are a Medical History of the Post; a Morning Report of Sick and Wounded; a Register of Patients; a Register of the Hospital Fund; a Register of the Physical Examination of Recruits; a Record of Deaths and Interments; an Order and Letter Book; a Meteorological Register at certain posts, and a Book of Information Slips for use when formal letters are unnecessary. In addition to these are the following

[1] Blanks from the Inspector General's Office.
[2] Blanks from the Quartermaster General's Office.

from the Adjutant General's office: A Descriptive and Deposit Book; a Morning Report Book of the Hospital Corps, and a Clothing Account Book.

19. Drills by word of command are needful to perfect men in movements that require concerted or co-operative action. It is a mistaken notion to suppose that because a drill is authorized and provided for, the various details of that drill must be rigidly observed on every occasion. The drill is merely a means to an end. A well-manned battery keeps up a rapid fire on the enemy because every man at every gun knows the duty devolving upon him, and does it without command at the precise moment when it should be done; but this perfection of co-operative work can be attained only by repeated and careful drills in the consecutive movements, each executed at the word of command. An analogous drill with the litter, ambulance, and a representative of the disabled human body familiarizes men with the management of these objects, and prepares them to act intelligently one with the other and irrespective of commands when the necessities of the occasion require such action.

20. The *Drill Regulations for the Hospital Corps*[1] prescribes the method of formation and alignment of the detachment, its marchings, turnings, rests, and dismissal. It then describes the litter and the methods of handling it, closed, open, and loaded. After these instructions come paragraphs on improvisation of litters; the removal of wounded without litters; the use of the travois, the horse, the two-horse litter, and the ambulance; inspection and muster, and the pitching and packing of hospital tents. The book closes with *Outlines of First Aid*, which are used as a text for the instruction of the enlisted men of the line, who

[1] Washington, D. C., Government Printing-Office, 1896.

are required to be drilled by their company officers for at least four hours in each month in the duties of litter-bearers and the methods of rendering first aid.

21. The *uniform* of the corps *for ordinary wear* consists of a dark blue flannel blouse and trousers of light blue kersey with stripes of emerald green, one half inch, one inch, and one and a half inches wide respectively for privates, acting stewards, and stewards. A brassard of white cloth, three inches wide, with a red cross in the centre is worn on the left arm of the private soldier. The chevrons of the acting steward, consist of three bars of emerald-green cloth with a red cross within, worn above the elbow, points down; those of the steward have in addition an arc of green cloth, over the ends of the bars. The waist-belt is of leather, black, with plate. The cap is of dark blue cloth, with a rounded and sloping visor of black patent leather, the private and acting steward having a white metal cross in front, and the steward a similar cross in a wreath of white metal. The equipment consists of the litter-sling and the hospital corps or orderly pouch.

The *uniform for ward service* is of cotton duck.

For field service the cap of the barrack suit is replaced by a campaign hat of drab-colored felt, and the legs are protected by leggings of strong cotton duck. The equipment for field service consists of the litter-sling, hospital corps or orderly pouch, canteen, and haversack.

Uniform for field service.

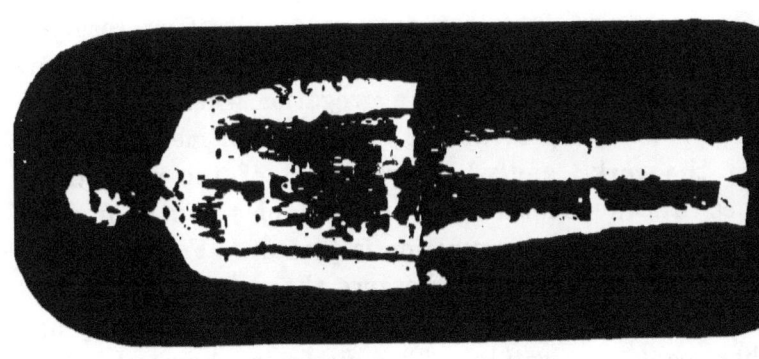

Uniform for ward service.

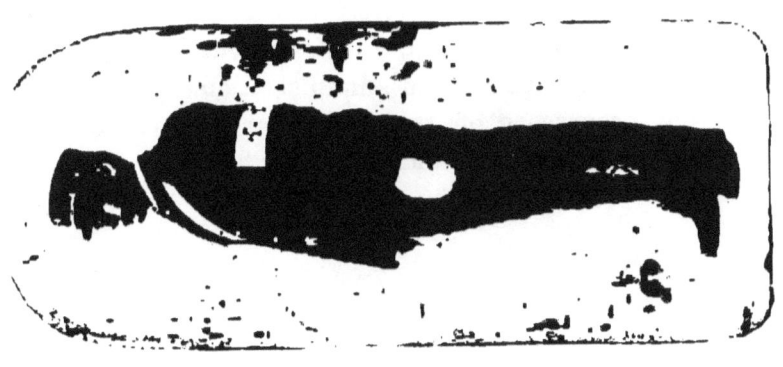

Uniform for ordinary wear.

CHAPTER II.

ACTIVE SERVICE IN THE FIELD.

22. The size of a hospital establishment is proportioned to that of the command to which it is attached; for, when troops are concentrating, every subordinate arrival brings its quota to the hospital corps. At first, in the progress of concentration, every material addition to the strength of the command necessitates a corresponding change in the organization of its hospital, to provide their share of the work to the newcomers. For organization consists essentially in so apportioning the work to be done by a number of individuals as to produce satisfactory co-operative results. But during the progressive concentration of troops there comes a stage or period of the aggregation when the hospital corps is able to provide one man or set of men for every part of the work; and when every part of the work thus assigned draws out the full energies of the man or men charged with it, the organization is perfect. Further additions do not strengthen it, because there is no room for them. They can be used to advantage only in building up another such organization. This perfected hospital is the unit of organization for field service in time of war; and the field hospital system of a large army consists of a series of such units, just as the army itself consists of a series of regiments which are the units of its organization for purposes of drill, discipline, and administration.

23. The occurrence of the Civil War first aroused the medical profession of this country to the necessity for a sat-

isfactory field hospital organization, but the succession of important events in this war was so rapid that practice had to meet emergencies without premeditation. In the end, success was achieved; but every step taken toward a better organization showed distinctly how much suffering would have been prevented or alleviated by an earlier recognition of its advantages. The lessons thus learned should never be forgotten; yet there is danger of their fading from memory unless they are treasured by the medical officers of the general and state governments as a precious heirloom bought with the blood and sufferings of their fathers for use in the possibilities of the future.

24. The unit of organization was at first the *regimental hospital*, but the inefficiency of this was speedily recognized. Its incompetency was strikingly conspicuous during the emergencies of the battlefield. The medical strength of the army was scattered along the rear of the line of battle. Some regiments suffered more than others, and their ambulances and stretcher-carriers were unable to remove the wounded promptly from the field; their medical officers were overworked, yet could not accomplish all that they desired; their shelters were insufficient and their supplies inadequate. Nor in this system could there be any efficient co-operation, for other medical officers, whose regiments perhaps had not become engaged, had to hold themselves in readiness for the developments of the battle; and although they might be ready and willing to assist their overworked comrades by their personal labors, they were warranted in showing some hesitancy in sharing with others the stores and dressings provided for their own men, if there was a probability of their own command becoming engaged before an opportunity would be afforded of replenishing supplies. So the better to provide for the emergencies of battle, a

temporary consolidation was directed whenever an engagement was imminent; and as the medical staff of a brigade was able to furnish an officer for every special duty connected with the hospital, and yet leave enough to give first aid on the battlefield, the organization was usually effected by brigades. In this way, when one regiment suffered more than another, the medical officers of several regiments participated in the care of its wounded, and individual cases requiring operative proceedings came under the hands of those surgeons in the brigade best qualified to undertake them. At this time, as soon as the surgical work of the engagement was completed the hospitals reverted to their regimental status. An active campaign, however, or a quick succession of battles, speedily demonstrated the advisability of retaining the brigade organization as long as the troops were within striking distance of the enemy; and while preserved in this way for a probable emergency, the consolidated field hospital for the brigade had an opportunity of showing its superiority to the regimental hospitals, as well during the marches and strategic manœuvres of active service as during its battles. Fewer wagons were required for the transportation of its property and supplies than for those of the four or five small hospitals which it replaced, because there was no unnecessary duplication of material; and the sick and wounded were held better in hand for sudden movements. Under the regimental system the sick of all the regiments were carried in ambulances, which generally followed in rear of the division on the march, and, at its close, were distributed among the several regimental camps. Here the sick had to await the arrival of the heavy trains before shelter or food could be provided for them; and it was precisely when both were most required, that is, during rainy and inclement weather, that the delay in their arrival was greatest. They were cared for in the ambulances during the day, but at night were transferred to their regimental camp, where their regimental hospital had merely an official, not an actual, existence. Under the brigade system the hospital camp was formed where the ambulances halted, and food and shelter were immediately provided, irrespective of the arrival of the main supply train. Regimental hospitals ultimately disappeared from the camps of veteran troops even during seasons of inactivity and recuperation, their official existence being represented merely by the regimental surgeons, who gave first aid in emergencies, and examined the command daily to find out who, if any, should be sent to the consolidated hospitals. A regimental hospital organization exists in most of the state military forces, as this system appears to answer the purposes of the Guard in their home service. Volunteer troops, tendered by the State for Federal service, would likely have a similar medical organization; but when brigaded for actual war service the field hospital system of the Civil War should be substituted.

THE DIVISION FIELD HOSPITAL.

25. But the brigade hospital, although a great advance on the regimental system, was found defective on many occasions when the brunt of the battle fell on a particular part of the line. A higher organization was found to be necessary; and this was effected by consolidating the brigade hospitals of the same division into a *division field hospital*. One medical officer exercised supervision over its various parts, the brigade hospitals becoming merely wards or sections of this larger organization. This consolidation, effected, it might be said, under fire on the field of battle,

continued to the end of the war to give thorough satisfaction to those who had most experience of the difficulties to be overcome.

26. At the post hospital the ambulance wagon is one of the belongings of the hospital, and when there is need for its use the driver and attendant litter-bearers are temporary assignments from the hospital force; and with small detachments in the field, it similarly forms part of the field hospital. But when the field hospital has reached that stage of its growth which requires the presence of a surgeon in charge for its proper management, the ambulance service must be placed under a special officer, for on important occasions the duties of the ambulance service lead it where the surgeon in charge cannot be present to superintend.

27. The members of the hospital corps on duty with the ambulances of the division are organized under the command of an officer, with brigade sections under junior officers espcially assigned, and with stewards and acting stewards as non-commissioned officers. The strength of the command is proportioned to that of the division, and is determined by the Surgeon General. For a division of 10,000 men, in three brigades, there should be three brigade sections aggregating 50 ambulances, 50 non-commissioned officers, and 250 privates. These would suffice to meet the requirements of an active campaign under all but the most unusual circumstances. The surgical history of our Civil War testifies to this. Improvements in firearms have changed the conditions since then, but as yet, fortunately, we have no other statistics to quote. The maximum percentage of loss which may befall a command depends upon the size of the command. Some company always suffers more than the average loss of the regiment, some regiment more than the average of the brigade, and so on. The

THE DIVISION FIELD HOSPITAL. 21

larger the command, the less the percentage of loss. Suppose that 1,500 men were struck down in the division of 10,000 men. The records of the war show 4.565 men wounded for every man killed. Of the 1,500 struck, 270 would remain for burial, while 1,230 would be entered on the list of wounded. The number of those who reach the hospital without assistance is always large, but it is relatively larger after severe engagements, for when the dressing and ambulance stations are crowded many will undertake the journey on foot rather than wait for the return of the wagons. Of 245,790 gunshot injuries recorded in the Surgical History of the War of the Rebellion, 56 per cent. were wounds of the upper part of the body and upper extremities, including fractures of the bones of the hand, which did not prevent their recipients from finding their way to the rear without the assistance of the ambulance service. This disposes of 689 of the 1,230 cases. Twenty-four per cent. were fractures of the upper extremities, flesh wounds of the lower extremities, and fractures of the bones of the foot which did not require operative procedure at the hospital. These cases, numbering 295 of the 1,230, may be regarded as having been able to bear transportation in the sitting posture. The remaining 20 cent., or 246 of the 1,230 cases, were of such a nature as to require carriage by ambulance or litter in the recumbent position. Viewing the capacity of an ambulance at seven, six inside and one with the driver, and allowing one recumbent passenger to be equivalent to three seated, the number of seats required for the transportation of the wounded in the division which lost 1,500 men in killed and wounded would be 1,033, or about three trips for each of the 50 ambulances. Of course, an ambulance may make its trip without being fully loaded, but as a rule, when the wounded are as plentiful as in the case supposed, there is seldom any spare space in a wagon which is starting from a dressing-station.

28. Existing orders do not provide specifically for the personnel of the Division Hospital. Its organization is left to the discretion of the chief surgeon. A steward, an acting steward, and ten men for each brigade, in addition to the teamsters of the supply wagons, are required for efficient service. When the hospital is crowded with wounded after a battle, details may be made from the ranks of the litter-bearers whose special duties for the time being have ceased.

29. The *surgeon in charge* is responsible for the care of the sick and wounded on the march and in camp, and for the comfort and general welfare of the wounded when brought to his establishment by the ambulance service. He makes requisition for medicines, medical and hospital stores, supplies, and property, and is responsible for their proper expenditure or use. An *executive officer* aids him in his work of supervision, and has special charge of the records. A *subsistence* officer superintends the cooking and diet of the hospital, drawing rations from the Subsistence Department, issuing to the brigade sections, and keeping the accounts of the hospital fund; he has also special charge of the hospital stores and of such articles of hospital property as are connected with the cooking and serving of food. A division steward looks after all articles of property borne on the returns of the surgeon in charge, taking care that by timely requisitions all deficiencies are made good and the hospital is always prepared for the coming emergency.

30. The *attending surgeon* cares for the sick of his brigade on the march and in camp, and during an engagement looks after the management of his wards, making notes of operative procedures, deaths, and of the progress of cases for subsequent report to the surgeon in charge and entry

on the records of the hospital. Each brigade section requires for its complement a steward and one or more acting stewards, the former to have charge of the medical supplies and instruments, the latter to act as wardmasters. In the wagon train should be carried from thirty to forty hospital tents, with picks, spades, and other implements for use in pitching the tents, trenching the ground, digging sinks, burying the dead, etc.; bedsacks and blankets, and cots, mattresses, and pillows for special cases. The surgeon in charge should endeavor to have cots or spare litters for all the cases that his canvas will cover, but restricted means of transportation generally prevent this. Woollen and light rubber blankets are brought from the field with the wounded, so that the hospital stock is made use of only in cold or wet weather, or to supply occasional necessities. Sheets, drawers, and socks should be provided to replace those that have become unfit for use. Rations of bread and beef stock, tea, coffee, sugar, and salt are carried for use during an engagement to insure food to the wounded until communication is opened with the main supply trains after the battle; and when this communication has been effected these stores should be immediately replaced to provide for the next emergency of the kind. The ambulances may also be utilized for the transportation of such supplies. Each should be fitted up with a water-keg and a locked box, the latter containing beef stock, tea, sugar, and hard bread, except in one instance in each brigade, in which the contents should consist of anæsthetics, morphine, antiseptics, and dressings. Each should also be provided with an axe, lantern, candles, and an iron pail which can be used in case of need as a camp-kettle. By this arrangement field supplies and stores are found by the side of the wounded as soon as they are brought from the field, and the work of the hospital can progress on this preliminary supply until the arrival of the main train. The heavy wagons of the hospital carry also the ordinary army ration for its employees and sick for the number of days that will elapse before new issues are made from the base or general supply train; the kitchen outfits; the tents and personal baggage of the officers; the blanket-rolls of the men, and forage for the horses and mules. The weight and bulk of this material are such that ten six-mule army wagons will suffice for its transportation. The capacity of an army wagon, on good roads and with full forage for the animals, is about 3,000 pounds. A blacksmith's forge accompanies the train.

ON THE MARCH.

31. Camp is usually broken up soon after *reveille*. Ambulance-drivers and teamsters groom, feed, and water their horses; litter-carriers pack up their shelters and blankets, and fill the ambulance kegs with fresh water, while nurses and cooks attend to their respective duties. After breakfast, the sick are examined and medicines prescribed and provided for their use during the day. They are then transferred to the ambulances, while the hospital tents and bedding, kitchen utensils, and other property are packed up and stowed away in the heavy wagons, which by this time have reported for their loads. The ambulances, medical wagons, and those carrying the subsistence of the hospital usually take position in the column of march immediately in rear of the division; but the transport wagons remain in camp until the troops of the army corps or of the army have passed, when they join the column of the regimental baggage, ordnance, subsistence, and forage wagons, under the protection of the rear guard of the army corps or army. Generally, these wagons come into camp at the close of the day's march shortly after the troops and their ambulances have reached it; but occasionally the conditions are such that the separation is for a longer period.

32. The ambulance train on the march receives those who have fallen out of the column from accident or disease. Usually each of these new cases has been examined by the regimental medical officer and furnished with a permit to await the passage of the train. The march has its sufferers as well as the battlefield. With raw troops, the sick and exhausted accumulate from day to day, until it becomes necessary to relieve the ambulances by sending the serious cases to the base of supplies. In the absence of transportation, this excess of sick may have to be left in extemporized quarters, or, if need be, in a section of the field hospital establishment.

33. At the conclusion of the day's march, when not immediately in front of the enemy, the ambulances are halted in rear of the division. While waiting the arrival of the baggage wagons, the sick are examined and treated. Such as are considered fit for duty are sent under charge of a steward to their regiments; and a notification is sent to regimental surgeons in the case of men admitted without permits and retained as unfit for duty, that these men may be properly accounted for on the regimental reports. Meanwhile, wood and water are procured, and the fires are lighted.

34. As soon as the baggage train arrives, the men on duty as pioneers and laborers unload and pitch as many hospital tents as will accommodate the sick, and afterward the tents of the officers; the nurses fit into the tents such bedding and other articles as are needful; and thereafter the litter-carriers transfer the sick from the ambulance wagons. Meanwhile, the cooks provide a refreshment of tea,

coffee, or consommé. Later in the evening dinner is served. Pending its preparation, the ambulances and baggage wagons are parked in rear of the tents, and the horses and mules fed, watered, and groomed, while the litter-carriers pitch their shelters between the wagon park and hospital tents, and the pioneers trench around the wards to keep their floors dry in case of rain. With dinner the labors of the day are at an end, save for the wagon guard, the hospital guard, if the season requires one to attend to the fires, and the special work of medical officers and nurses in particular cases. In the establishment of this camp each man, by drill and experience, knows his particular duty, and by doing it well enables the whole to be accomplished with ease and rapidity. Less than an hour will suffice to transform a deserted field into a hospital settlement as orderly and perfect in its field appointments as if it had been in existence there for many days.

ON THE BATTLEFIELD.

35. To meet the emergencies of battle, a standing order is issued the requirements of which take effect as soon as it is evident that a struggle is imminent. This order prescribes the duties of each medical officer during the anticipated engagement. An operator and assistants report to the surgeon in charge of the hospital for duty in each brigade section. The other medical officers are assigned to duties on the field or at the hospital, as may be determined by the chief surgeon. A severe engagement seldom takes place without premonitory signs. The chief surgeon is aware of what is impending, and satisfies himself that his command is well in hand. As soon as he learns the position of the line of battle, he indicates to the surgeon in charge, personally or by messenger, his views as to the location of

the hospital. The ambulance officer, on his return from a survey of the roads leading to the front, may be the bearer of this information. The particular locality in this neighborhood is selected by the surgeon in charge, with due consideration to questions of water and fuel, dryness of site, facility of communication with main roads, and availability of neighboring buildings as hospital accessories.

36. The hospital should not be too near the front. Nothing is so depressing to the wounded, already more or less prostrated by their injuries, than exposure to fire while under the hospital flag, as it is suggestive of a disaster to the line of battle. Even in the best-disciplined establishments the effect is sometimes demoralizing. The hospital cannot be moved farther to the rear under these circumstances without detriment to the wounded already brought in, while those on the field would have to be left for so much longer before obtaining shelter and care. Nor, for manifest reasons, should the hospital camp be too far from the front. A distance of from one and a half to two miles will give fair security; while, if the roads are good, there will be little delay in the transport of the wounded. The immediate vicinity of a farmhouse, country seat, or other dwelling affords many advantages. Water, fuel, and direct communication with the main roads are usually associated with it; it offers a point of prominence for the display of the hospital flag, should it become exposed to the fire of the enemy; and its rooms may be made use of as adjuncts to the hospital should the number of wounded exceed the capacity of the tents to accommodate them.

37. If the hospital train is not on the ground, special messengers are despatched to hasten and guide it, and on its arrival the pioneers, nurses, cooks, and teamsters proceed with the routine work of unloading, pitching, trench-

ing, and furnishing the tents, building fires, and preparing for the issue of beef soup, tea, and coffee, every section or part of the work under the personal supervision of an officer, every squad controlled by a non-commissioned officer, and every man familiarized by practice with the duties required of him. At the same time, on the ground allotted by the surgeon in charge, each brigade steward opens his wagon for service, and has the operating fly of his command pitched and furnished with its table, instruments, appliances, and supplies pending the arrival of the operating staff from the front.

38. On orders issued by the Adjutant General of the Division the services of musicians and members of the drum-corps may be utilized in pitching tents and preparing the hospital camp for the reception of the wounded. Thereafter they may be employed according to their individual capacities, the main body, however, being assigned to duty as the police party of the camp. Some medical officers have reported against the use of drum-corps details for such duties on the ground that they are troublesome and unmanageable; others, however, have credited them with valuable services.

39. Meanwhile, at the front, the chief surgeon supervises the arrangements for first aid to the wounded. Existing orders require that the wounded shall receive attention at three points on their way to the division hospital: 1st, on the line of battle; 2d, at the first dressing-stations; and 3d, at the ambulance stations.

40. The attention they receive on the line is that afforded by the litter-bearers of the ambulance company, who have reached the front, under command of their company officers and stewards. First aid here is limited to giving water or some restorative, relieving the wounded man from harmful pressure of equipments, and getting him by the shortest and safest route to the dressing-station. Dangerous bleeding is controlled by pressure and fractured limbs are adjusted for transportation on the litter.

41. The dressing-stations are situated at the nearest point where protection may be obtained from musketry fire. If the men are fighting behind breastworks, the best protection may be on the line itself; otherwise, advantage is taken of some superficial depression, gully, ravine, fence, wall, or building, two or three hundred yards in rear. Under such cover as he can find, each medical officer who has been assigned to this duty takes position with a steward and orderly. The amount of surgical work performed here is greatly affected by circumstances. If the cases are numerous or the station exposed, many may be permitted to pass to the ambulances after a glance at their condition and a caution to permit no unauthorized handling of the wound, while the attention is mainly devoted to arresting bleeding, removing shock, and supporting fractures in slings or light splints for ease in transportation. But if casualties are infrequent and the station well protected, flesh wounds may be thoroughly cleaned with boiled water, covered with bichlorid dressings, and the patient tagged to intimate that further interference is unnecessary. Dressing-stations are distinguished during the day by Red Cross flags and at night by red lanterns.

42. The ambulance stations are situated as close to the rear of each brigade as the nature of the ground will permit. Here the medical officers clean and place protective dressings on wounds that escaped attention at the first dressing-stations, marking such as require no further investigation, and also such as seem to call for immediate operative procedure. They superintend the loading of the ambulances, and see that every case is in the best possible condition to

undertake the journey to the hospital. The topographical features of the battlefield are often such that the first dressing-stations and ambulance stations may be consolidated. When the ambulances can get close up, there is no need for an intermediate halt in the removal of the wounded; and when the roads and the ground permit of it, ambulance stations should be established at more than one point in rear of a brigade, in order to shorten the distance over which the wounded have to be assisted or carried. In the event of an advance or a yielding of the line of battle, a corresponding change of position of the dressing and ambulance stations must be effected. In the former case, a few of the ambulances follow up the advancing line, leaving the greater part to clear the field before participating in the forward movement. In the latter case, positions are assumed in rear of the re-formed line, except when the hospital becomes exposed, in which case special orders from the chief surgeon determine the further movements.

43. Generally, some time before the first ambulance load of wounded arrives from the front, the surgeons on duty at the hospital are engaged in receiving, dressing, or operating on, injuries of the hand and flesh wounds attended with little hemorrhage or shock. These cases probably left the field stations without waiting for attention and made the journey unaided and on foot. Each is assigned to a specified ward, the acting steward or wardmaster of which is thereafter responsible for his comfort. Shelter-tents are pitched at regular intervals near the hospital tents to form the ward for these lighter cases. The pioneers attend to this work, trenching the ground, weather-guarding the open ends of the shelters, and providing some material, as hay, straw, freshly cut grass, leafy twigs, wood-shavings, etc., for bedding. If no suitable material can be found in the immediate vicinity, one of the now-empty transport wagons may be despatched to some point where a supply may be obtained; and if no such point is known, a detachment of the drum-corps, under responsible leadership, may be sent out as foragers. During inclement weather these slighter cases may be housed in the as yet unoccupied hospital tents until their special camp is prepared. When thus systematically camped, their wants are not overlooked, as each wardmaster has his duties aggregated and defined.

44. When ambulances arrive from the field, the whole staff of the hospital becomes at once engaged. Should the reports from the front indicate that the tents will be insufficient for the accommodation of the wounded, the flies are moved forward to extend the wards, and extra bedsacks are filled with such material as may have been collected. If this extension is insufficient for the shelter of the in-coming wounded, the neighboring dwelling or its outhouses may be utilized.

45. The end and aim of the work of the hospital is to have all primary operations completed and the wounded ready for transportation to the base of supplies at the earliest possible moment, because an advance or retreat is sure to follow the battle unless it has left both sides unfit for immediate aggression. The important part of this work falls upon the operating staff. The surgeon in charge should therefore see that these officers are provided with a steady succession of cases until the work is finished. Time lost in field surgery is lost between cases for want of that systematic direction which enables all to be constantly employed.

46. The chief surgeon provides the transportation for the removal of the wounded from the hospital. If railroad or steamboat communications are available, the ambulances of

the division may be used to convey the sick and wounded to the station or wharf; but if the journey is one of considerable length, these wagons should not be employed unless they can return to the hospital in time to secure needful rest before the march is renewed. Ambulances from the base may sometimes be sent forward to relieve the hospital. Usually, however, the empty wagons of the subsistence and ordnance trains are employed. The surgeon in charge superintends the loading of these wagons, and provides the medical officers to accompany them, with such articles of food and medicine as are needful for the journey. He may have to part with many of his mattresses, bedsacks, and blankets in outfitting the train, but requisitions to replace them, and to replenish supplies generally, may at the same time be transmitted to the purveyor at the base.

47. The medical officer in charge of the wagon train should be provided with a nominal list of the sick and wounded intrusted to his care. Frequently, however, the military conditions are such that no list of names can be made out. At starting he may be able to learn of his train merely that it consists of so many medical officers, or, if small, of so many stewards and acting stewards, each of whom reports himself responsible for so many men and their supplies. On delivering up his charges to the authorities at the base hospitals, he should report back to the surgeon in charge of the division hospital all the deaths or other losses that have occurred during the journey, that they may be entered on the register and communicated through regimental surgeons to the company officers on whose muster-rolls the names of the men in question are carried.

48. If, as an immediate result of the battle, the enemy withdraws to another position, it is not necessary for the hospital to follow up the advance of the troops unless the distance is considerable. Should this be so, the hospital moves forward, carrying the wounded, if few in number, with it, or leaving them in a detached section of the establishment until they have been put in condition to undertake a rearward journey in the supply wagons.

49. Should the battle be indecisive, the losses will probably be great, but time will usually be afforded for the completion of the surgical work, because neither party is in a condition to renew the contest. The hospital, therefore, remains undisturbed probably for several days; but so great is the uncertainty of battlefield conditions that every effort should be made to complete the surgical work. Promptly on the cessation of the battle, medical officers who have been on duty at the stations report to the surgeon in charge for assignment to temporary duty at the hospital. Extra surgical help may also be drawn from the base or general hospitals, if telegraphic and railroad connections have been kept up.

50. Should the troops in the line of battle be driven back, the exposure of the hospital is unavoidable, unless anticipated by prompt action on the part of those in charge. The hospital and its accumulated wounded should be moved to a suitable site in rear of the new position. The wounded left upon the field must be cared for by the medical department of the opposing force; but such as have reached the shelters prepared by friends should not be given up without the strongest efforts to save them. Medical officers and members of the hospital corps should be officially detached to remain with any left behind; and shelter, bedding, medical and hospital supplies, and food should be amply provided for them. Tents and supplies thus lost should be replaced by immediate requisitions on the supply depots.

51. If the disaster is so serious that none of the wounded can be removed, the surgeon in charge and ambulance officer should endeavor to preserve the organization by withdrawing the ambulances, wagons, supplies, stores, and personnel not specially assigned to remain with the wounded. When the hospital is thus disabled, an immediate renewal of hostilities necessitates the occupation of the available buildings in the vicinity of the new site; but this utilization of pre-existing shelter would have been necessary if, without the repulse, the wounded had been equally numerous.

52. If the disaster involves the capture of the transportation and supplies, while certain of the officers and men have escaped, these must be organized by the senior medical officer for the duty of caring for the wounded who may fall in the skirmishes of a subsequent retreat. Notwithstanding breaks in the ranks, their training holds them together for this special service until the command is re-enforced and refitted.

53. When a regiment or brigade is detached for permanent assignment to some other command, it carries with it, when needful, its section of the ambulance company, hospital property, and transportation; but when this detached command is en route to a rendezvous where it can be refitted, its section of the hospital is left behind and applied by the chief surgeon in repairing deficiencies in other brigades, or is held subject to disposal by higher authority.

54. The Treaty of Geneva has of late years done much to mitigate the sufferings of the wounded of a defeated army. The Articles of the Convention, which were adopted by every European power and the majority of the South American States, at different times since 1864, provide:—1st. For the neutrality of ambulances on the battlefield, and military hospitals as long as they contain any sick; 2d. For the neutrality of the staff, medical, and administrative officers, attendants and litter-bearers; 3d. That the neutrality of

these persons should continue after the occupation of their hospitals by the enemy, so that they may stay or depart as they choose; 4th. That if they depart they may take only their private property with them, except in the case of ambulances, which they may remove entire; 5th. That a sick soldier in a house shall be regarded as a protection to it, entitling the occupant to exemption from quartering of troops and from part of the war requisitions; 6th. That wounded men shall, when cured, be sent back to their own country on condition of not bearing arms during the rest of the war; 7th. That hospitals and ambulances shall carry, in addition to the flag of their nation, a distinctive and uniform flag having a red cross on a white ground, and that their staff shall wear an arm badge of the same colors, the delivery of which shall be left to the military authorities; 8th. That the details shall be left to the commanders of the armies. The 9th and 10th articles are formal and signatory.

55. In 1868 supplementary articles were agreed upon, but they have not been ratified, though they were practically adopted by Germany and France in the war of 1870. Their provisions affecting armies are that, when a person engaged in an ambulance or hospital occupied by the enemy desires to depart, the commander-in-chief shall fix the time for his departure; and that, if he remain, he shall be paid his full salary; that the ambulances mentioned in the 1st and 4th articles include field hospitals; that in exacting war requisitions account shall be taken not only of actual lodging of wounded men, but of any display of charity toward them; and that the rule which permits sound soldiers to return home on condition of not serving again shall not apply to officers, as their knowledge might be useful.

56. Officially there is only one Red Cross, that of the

Geneva Convention; but many good people in certain countries, when the treaty was signed, formed themselves into Red Cross societies to collect funds and stores for relief purposes when war should come; and these societies assumed the Red Cross as their flag and insignia. They have, however, no legal or official status in connection with the army, and are entitled only to such privileges as the military authorities may desire to extend to them in the interest of the sick and wounded.

57. When the army goes into a comparatively permanent camp, as in winter quarters, during sieges, or in the occupation of hostile territory, the sick and wounded need not be sent away unless they accumulate beyond the capacity of the hospital to accommodate them. Commanding officers generally approve of retaining the men in the field hospital, as return to duty on recovery is better assured than from distant hospitals. They therefore further the efforts of the medical department in improving the condition of the hospital. Lumber is obtained, and the tents are framed, floored, and weather-boarded, while shelves, tables, and benches are put in, with suitable stoves or brick fireplaces to warm the wards. Board walks are laid, and the grounds fenced in and thoroughly policed. As thus established the field hospital presents an air of permanence and stability; but it should be kept in proper drill, and ready at a few hours' notice to be packed up and following the division in the column of march.

OFFICIAL PAPERS.

58. The medical records of field service in time of war have an importance which is not always recognized by those who are responsible for their accuracy and completeness. While facing the suffering of the battlefield the mind be-

comes careless of prospective considerations, and sometimes looks upon the preparation of reports and papers as akin to substituting the so-called red tape of bureaucratic methods for the antiseptic dressings of the practical surgeon; but the value of these records may be at once appreciated when it is realized that they are the corner-stone on which the pension system is built. Defective records may cause much suffering in the future by delaying or preventing the establishment of just claims for relief.

59. The senior medical officer of each regiment co-operates with company officers in providing the regimental commander with the materials for his *Field Return of killed, wounded, and missing*, which is filed in the office of the Adjutant General, as the official record of the losses. The senior medical officer is called upon also to forward to the chief surgeon duplicate lists of wounded within two days after an engagement. Outside of these battle returns the only reports required from this officer are the *morning and monthly reports of sick and wounded* of his regiment. He reports those that are in the field hospital as well as those who are excused from duty on account of trivial ailments. His morning report is for the regimental commander, but a duplicate is sent to the chief surgeon. This report should give also the names of such hospital corps men as are on detached service with the regiment. In large commands, as an army corps, a *weekly report* may suffice to keep the chief surgeon informed as to the condition of the troops; but the medical officer on the staff of generals of brigade or division may call for a copy of the morning report. The monthly report forms a part of the permanent record, and is transmitted to the Surgeon General.

60. The surgeon in charge of the hospital reports his sick and wounded daily to the chief surgeon for comparison with

the reports from the regiments; and monthly to the Surgeon General and chief surgeon. He is accountable for the medical and hospital property and supplies in use, signing all requisitions, and making annual returns. He reports the personnel of the hospital corps daily to the chief surgeon, and monthly to the Surgeon General and chief surgeon. As commanding officer of the hospital corps, he keeps the accounts of the pay, clothing, etc., of its members, including their final statements in case of discharge or death, the executive officer relieving him of the details of these duties. He is responsible also for the subsistence of his hospital and for the proper expenditure of its hospital fund, the subsistence officer aiding him in these duties.

61. The ambulance officer is charged with the care of the pay, clothing, and subsistence accounts of his men, and is held responsible for the care of the ambulances, wagons, tents, horses, mules, forage, etc.

CHAPTER III.

SANITARY CARE OF CAMPS.

62. The site of a camp is of the first importance, because, in the event of its insalubrity, no exercise of care in the sanitary government can protect from evil consequences. Ordinarily, in selecting a site the health and comfort of the men is the first consideration; but when a military object is in view, the ground must be selected and the camp arranged for that object, other considerations being merely secondary.

63. Dryness of site is essential to the healthfulness of a camp ground. It depends on the inclination of the surface and the porosity and depth of the subsoil. The *natural drainage* of a place is good when the surface sheds the rainfall into neighboring watercourses, or when the surface layers are so porous as to soak up the rainfall and drain it off to lower levels by underground channels. The natural drainage is bad, and the site damp and insalubrious, when a level or slightly undulating clayey surface retains the rainfall in shallow pools, or where the level of the *subsoil water* is near the surface. The subsoil water is that which is found in digging shallow wells — rain-water which, having penetrated a porous surface, is upheld below by a stratum of impervious clay.

64. Moist soils, under their most favorable aspects, induce catarrhs, sore throat, and other internal inflammations, develop consumptive and rheumatic tendencies, and, by depressing the vitality of the system, render it an easier prey to the attacks of other diseases; they also give rise to an influence known as *malaria*, which is recognized as the cause of intermittent, remittent, and congestive fevers, enlargement of the spleen, congestion of the liver, dysentery, many neuralgic affections, and that broken-down condition of the system found in individuals who have lived for some time in marshy districts.

65. Malarial fevers are caused by the presence of a microscopic parasite, the Plasmodium malariæ, in the blood. The life history of the plasmodium outside the body is not known, but it is believed to be connected with the fermentation of the organic matter in soils. For the production of malarial manifestations there are needful: 1st, a certain amount of decaying organic matter; 2d, a certain proportion of moisture; and 3d, a certain degree of heat. Under the influence of heat and moisture, the organic matter undergoes a fermentation during which the pernicious element is evolved. If organic matter be not present, there can be no fermentation; excess of moisture retards the fermentation by reducing the heat below the required temperature; and excess of heat produces the same effect by dissipating the needful proportion of moisture.

66. This theory fits well with much of our knowledge of malaria. Dams, lakes, and ponds with an equable water-level and well-defined margins are not unhealthy, but grounds that are alternately submerged and exposed are malarious. The artificial draining of ponds and the accidental breaking of dams have occasioned disease in their vicinity. Fever and ague prevail not during periods of inundation, but during the subsequent periods of draining and drying up, if the season be sufficiently warm. Shallow mill-dams that uncover a portion of their storage area during the use of the water are more dangerous than those that have depth enough to keep the area submerged. Shallow creeks which open into salt water, and are subject to tidal influences, are generally malarious.

67. Malaria is diffused into the atmosphere with some difficulty. When it drifts with the wind its course is along the ground; hence, in an unhealthy locality, it is more dangerous to sleep on the ground or on the ground floor of a house than on a higher level. It is often associated with mists or fogs, which, hanging low over the exhaling surface, may be floated into neighboring valleys and upward along the rising grounds. Floating mists of humidity and malaria are intercepted by belts of trees: a growth of trees between a pestilent marsh and a settlement has frequently protected the latter from disease; and, conversely, the removal of a screen of trees has been followed by an invasion of malaria from neighboring swamps.

68. Although most of the recorded observations on the attributes of malaria are explicable by the theory of an exhalation from the soil during its fermentation, there are some exceptional points. When organic matter and moisture are present, the evolution of malaria appears to be proportioned to the degree of heat, yet it is well known that places which are deadly after sunset may be traversed with impunity when the sun is high in the heavens; and, although its evolution is known, by the experience of ages, to be most active in warm climates and in warm seasons, yet in a temperate climate the malarial influence does not become intensified during the hot summer months; but, on the contrary, after its first manifestations in the spring, it appears to lose much of its activity, until its autumnal period of greatest prevalence and virulence is suddenly reached.

69. If, however, it be assumed that the emanation from the soil be intended, like the carbon dioxid of the atmos-

phere, for the support of vegetable life, and that it manifests itself as a poison to animal life only when it is generated and exhaled in excess of the wants of the vegetation of the locality, the whole of the observed facts may be understood. During the day the vital activities of a luxuriant vegetation absorb emanations which, during the night, when vegetation is asleep, escape into the atmosphere as a harmful agency. When spring advances and warm rains fall, the fermentation of the organic matter of the soil begins immediately; but as yet there is no green foliage to absorb the emanations, and, in localities where the conditions for fermentation are particularly favorable, malarial fevers are found to be present. As the summer advances and the annual growth becomes vigorous and luxuriant, these spring fevers decline in prevalence. The luxuriance of vegetation on a particular soil indicates the presence of a malaria which would manifest itself by its action on the human system but for the existence of the vegetation. Later in the season the leaves wilt and fade, the seeds fall, and the plants droop and die; but meanwhile, under a continuance of the autumnal heat, fermentation goes on in the soil, and the evolved malaria, no longer absorbed or destroyed by the living vegetation, accumulates to a pestilential atmosphere.

70. While the *generation* of malaria depends upon fermentation, its *evolution*, so as to produce morbific effects on the human system, is due to a want of relationship between the growing plants and the malaria generated. When the verdure covering the soil is able to assimilate or dispose of the whole of the generated malaria, there is no evolution. If the fermentation in the soil be inactive, a covering of grass may suffice to suppress exhalation; but if the conditions are strongly conducive to fermentative action, there will be a richer growth of green vegetation on the surface, as in the tangled undergrowth of unhealthy tropical regions.

71. When an exact relationship exists naturally between the conditions of the soil and the vegetation covering it, there is no harmful evolution; but when that relationship is disturbed, malaria appears. The upturning of the soil for agricultural purposes—destroying existing vegetation for the sake of a future cultivated growth—is well known to give rise to malarial diseases; and even when continued cultivation has resulted in a due adjustment of the soil to its growing crop, popular experience has attached a baneful influence to exposure to the "night air" during the harvesting period. A similar result follows other interferences with the growing vegetation, as in engineering operations and the clearing of forests; even drainage for sanitary purposes is often harmful at first, by establishing new conditions as to moisture, which interfere with the luxuriance of the natural growth before a vegetation has been developed suited to the dryer soil.

72. From what has been said concerning malaria the value of certain suggestions in relation to camp sites may easily be understood: Avoid the neighborhood of marshes, river-bottoms, overflowed lands, deep alluvium, lands subject to occasional salt-water inundation, and sands with subjacent water, however barren their surface. If water cannot be had except in such localities, it is better to carry the water some distance than to camp in its vicinity. Grassy surfaces are usually accepted as good camping-grounds: the elimination is small during the night, lies low, and is completely absorbed in the early morning; were the elimination greater than is sufficient for the grass, other and more luxuriant plants would be growing on the soil. Shrubs on a moist soil indicate dangerous ground, for their

more extensive verdure implies an increased evolution which has to rise higher before reaching the absorbing surfaces; and, moreover, camp cannot be established without some clearing of the ground and the consequent diffusion of emanations which would otherwise have been absorbed.

73. There are other points which enter into the consideration of camp sites. Advantage in cold climates should be taken of hills and woods as a protection against wintry winds; and in hot climates, of woods for shade, if not so dense as to interfere with ventilation; prevailing winds should be observed, that the camp may be placed to windward of swamps or other insanitary localities. In the mountain districts of warm latitudes the nightly breeze from hill to plain must be remembered in its bearing on the healthfulness of sites. Cañons are hot during the day, oppressive at night by radiation from heated rocks, and liable to inundation from rain-clouds on the mountains; the reflected glare from sand and rock is often distressing and injurious to the men. A dusty site is hurtful to the eyes, combined, as it usually is, with a garish light; moreover, dust, like mud, renders the best-disciplined troops careless of their personal appearance and weakens the hygienic government. Old camping-grounds should be avoided on account of their filthy condition and the possibility of their infection [79].

74. Experience has shown that troops may be aggregated in camps on a healthy site without the occurrence of disease among them. Camp diseases are therefore preventable diseases. Among the diseases usually regarded as camp diseases are:

75. (a) Those occasioned by *exposure to climatic or meteorological influences,* such as catarrhs, bronchitis, inflammation of the tonsils, larynx, or lungs, rheumatism, con-

gestion of the spinal membranes, simulating to some extent rheumatic troubles, congestion of the bowels leading to diarrhea and dysentery, ophthalmia, sunstroke, etc. Inadequate clothing and shelter induce a greater prevalence of these morbid conditions among troops on active service in the field than among men surrounded by the comparative comforts of civil life; and the attacks are generally of greater severity on account of the more intense action of the cause and the sometimes exhausted condition of the troops, but in other respects there is nothing special in these diseases when they are found in a military camp.

76. (*b*) *Errors of diet*, attributable to faulty cooking, individual indulgence, imperfect mastication, and improper food are prolific sources of *intestinal irritation* which may end in dysenteric ulceration. The last includes all articles which have suffered damage by imperfect preservation, and meat which, while on the hoof, has been overdriven, badly fed, or affected with disease, or which has been kept too long in the slaughter-house after killing, or in the haversack after issue or cooking; and to these must be added many of the articles that are sold by sutlers and traders. Organic impurities in the drinking-water, and an excess of those salts which give hardness to a water, are occasionally responsible for diarrheal attacks. When the internal congestions which result from the malarial infection involve the intestinal mucous membrane, diarrheas are produced. Foul odors, as from unpoliced sinks or unburied carcasses, also occasion diarrheal efforts to rid the system of the noxious matters which have been absorbed from the air. Moreover, any influence which interferes with the normal action of the skin, as rapid cooling after cessation of exercise, may be a cause of diarrhea.

77. (*c*) Diet deficient in quantity predisposes the soldier

to disease by lowering the resisting powers of his system. When defective in quality from a sameness involving a deficiency of the salts that are contained in fresh vegetables and acid fruits, a taint of *scurvy* becomes manifest. This shows itself first in loss of spirits and disinclination for exertion, muscular pains simulating rheumatism, a slight tumefaction of the gums where they embrace the teeth, a slight fetor of the breath, and, perhaps, when specially looked for, some small spots of ecchymosis, like flea-bites, on the skin of the calf or other parts of the lower limbs. In all commands that have been confined to a salt ration for some time, these symptoms should be carefully looked for and promptly suppressed by an improved dietary. When the possibility of scurvy is not held in view, such cases are liable to be confounded with muscular rheumatism or diarrhea, for the latter disease, occurring in a scorbutic patient, is persistent and may for a time be held accountable for the deteriorated condition. Later, when the gums become swollen, spongy, and bleeding, the teeth loose, and the skin covered with ecchymosed patches, hard swellings, and foul ulcerations, there is no doubt of the character of the cases; but the disease should not be permitted to give such manifestations of its existence in a military camp. Camp hygiene requires the absolute banishment of alcoholic liquors from the lines. The medical and court-martial records of all camps where whiskey could be procured furnish data sufficient for insistance on its exclusion as the cause of much disease and many injuries and violent deaths.

78. (*d*) *Malaria.* Protection is obtained by care in the selection of sites, the avoidance of all unnecessary fatigue, and, in cases of special exposure, the use of prophylactic doses of quinine, with hot coffee and generous diet.

79. (*e*) The infection of *typhoid fever* is spread chiefly by the intestinal excreta of the patient; but many circumstances observed in connection with the development of this fever indicate that the evolution of its germs is associated with a fermentation in the soil when the soil has lost that proportion of moisture which is needful to the generation of malaria. Typhoid fever follows malarial fevers when man begins to drain for agricultural and building purposes. In some localities, which seem to be a border-land for the two diseases, malarial fevers prevail in moist, and typhoid in dry, seasons; and in America, as in Europe, the prevalence of typhoid has been found to coincide with a lowering of the level of the subsoil water and a corresponding dryness of the superficial strata. The facts and arguments in favor of a soil origin of typhoid fever, were they even less convincing than they are, would have to be accepted by the military sanitary officer as offering the only sure basis on which to attempt the preservation of his command from the attacks of this disease; for soldiers, unless prevented by active interference and supervision, will build quarters for themselves which are, to all intents and purposes, hotbeds for the production of fever [95], and the experience of every army has shown that the disease is seldom absent from such quarters when occupied by raw troops.

80. Every regiment of new troops in time of war undergoes a typhoidal seasoning, not necessarily, but because of the difficulty of excluding the cause of the disease. It is a specific disease, affecting the individual but once, as in that one attack it exhausts his susceptibility to its deleterious influence. Every new regiment, particularly if raised in healthy country districts, contains young men who are susceptible to the disease, and their number gives a corresponding susceptibility to the regiment. Such a command will suffer more from its localized epidemic of typhoid than

one raised in the slums of a city in which the disease is constantly present, for the majority of the young men from the unhealthy city have already undergone their experience of typhoid fever. The removal to hospital of each case as it occurs, the disinfection of excreta and soiled clothing [558], and the protection of the water supply will guard against the propagation of infection from that particular case; but if the disease be due to infection from the soil, these measures will be ineffectual. In fact, in view of the difficulty of preserving a susceptible regiment from succumbing temporarily to a typhoid epidemic, the medical sanitarian cannot afford to overlook the probable soil origin of the disease; and the measures he suggests for preserving the purity of the soil, water, and air of the camp and quarters of the men must be all the more thorough since the precise conditions of typhoid evolution have not been determined. Removal to a new site is always in order as a means of lessening the spread of miasmatic-contagious diseases.

81. (*f*) *Dysentery*, occurring as an epidemic in camp, is usually a miasmatic-contagious disease. It so frequently follows the breaking up of fresh ground that its miasmatic origin is unquestioned; and foul and infected sinks are supposed to spread the disease. Although it is difficult to prove the latter mode of propagation because its effects cannot be separated from those of the primary miasm, disinfection of the sinks and of all dysenteric discharges [536] is called for as a precautionary measure irrespective of theoretical views.

82. (*g*) *Cholera* and *yellow fever* cannot be considered camp **dis**eases, as they are as prone to attack the civilian as the soldier. In fact, the camp, being mobile and under better discipline, may sometimes be preserved while neighboring cities are prostrated by disease [559].

83. (*h*) The systematic vaccination of recruits at the camps of instruction and organization removes *small-pox* from the list of camp diseases. When an occasional case occurs the civilian employees of the Quartermaster's Department, and other attachés and followers of the army, are more likely to become affected than the fighting force [546].

84. (*i*) With *measles* [556], however, the case is different; there is no protection except that given by a previous attack, and when a regiment consists of susceptible material—that is, of young men who have not had the disease—its efficiency may be utterly destroyed for two months or more by the introduction of the infection. Individually the cases as a rule are by no means dangerous to life, but the conditions of service in the field are such that a satisfactory convalescence is impossible, and many men have subsequently to be discharged on account of persisting pulmonary complaints. Under such circumstances the regiment should be relieved from duty and provided with comfortable quarters and ample hospital accommodation until it has recovered from its attack. In this way only can it be saved from the deadly inroads of pulmonary disease, which are sure to follow the eruptive fever when its convalescents are subjected to the hardships and exposures incident to service in the field.

85. (*k*) *Typhus fever* was once the scourge of military camps and of all other places in which men were closely crowded together. Some of the names applied to it, *camp, ship, prison,* and *hospital fever,* indicate its association with overcrowding. But of late years the sanitary knowledge which has insisted on a certain air-space for each individual and a certain ration of air to revivify the blood in his lungs [230] has done much to stamp out its contagion. The experience of recent wars has demonstrated that typhus fever

can be generated at will, and that, if this fever *is* generated in a military camp, it is due to gross negligence or ignorance on the part of those in authority, for the disease does not spring unexpectedly into existence full armed for destruction, but gives successive warnings of its coming, each more emphatic than that which came before. When a number of men are confined in a limited and poorly ventilated shelter, the organic emanations permeate and adhere to everything, and continually taint the air. Overcrowding of this kind is incompatible with cleanliness even where facilities for cleanliness exist, but in such cases these are generally absent; and in consequence the fermentation of extraneous filth usually combines with the natural exhalations to alter the quality of the air and the constitutions of those who have to breathe it. The inmates lose vigor and appetite; they are subject to digestive disorders and headaches, and their sleep is unrefreshing. They become less able to withstand the ordinary exposures of service, and the febrile action which is associated with their bronchitis or other local disease is of an obscure yet prostrating character, as if the system did not have vitality enough to react and throw off the disease in a free perspiration, or a transudation from the affected mucous membrane. Presently, pneumonia appears to be the only disease that is developed by exposure, and this pneumonia is of that low or asthenic character which is called *typhoid*. If malarial fevers are present they are never well-defined and vigorous intermittents, but obscure remittents that are doubtfully treated with quinine—they have so much the appearance of true typhoid. If the command is suffering from an epidemic of typhoid fever the mortality rate is exceedingly high, as the patients sink into a state of prostration from which it is impossible to rally them. Moreover, at such times cases of sudden death in the persons of those who have not heretofore been ailing occasionally cause alarm in the camp. These are apparently due to some cause which deranges the blood, producing internal congestions, and perhaps cutaneous ecchymotic spots. They are called *congestive fevers* if the camp is malarious; *malignant measles* if that disease is prevailing; and *virulent typhoid* if the regiment is undergoing its typhoid seasoning; while the presence of epidemic *cerebrospinal meningitis* or *spotted fever* is feared by some, and *typhus fever* by others. At such times also there is an alarming suggestion of contagion in the hospitals and quarters, even in diseases which, like pneumonia, are not ordinarily regarded as having any contagious qualities. Many regiments during our late war suffered from disabling and needless experiences of this kind from an utter want of knowledge of the principles of camp sanitation on the part of those who should have possessed that knowledge. Fortunately this generation of the contagious typhus miasm was checked at an early period. The condition of the command attracted the attention of superior authority, and some officer of experience, sent as an inspector, appreciated the typhus-generating conditions to which the men were subjected, and speedily effected their removal. Some of these regiments were hutted in small squads of three to five men under shelter- [88] or wedge-tent [89] roofs, others were aggregated in larger squads in the Sibley tent [90], and others again quartered by companies in extemporized barracks or in rude buildings specially erected for their use; but in all such instances there was an utter disregard of the necessity for ventilation and cleanliness in addition to the overcrowding of the men. The tent-covered huts were usually *dugouts* [95]; the extemporized quarters wholly unfit for crowded occupation, and the barrack-buildings hastily constructed shells, enclosing three-tiered rows of beds, with no special provision for the introduction of fresh air, and all the crevices of their imperfect construction carefully stopped up by the men.

86. It will be seen from this brief review of the diseases which prevail among soldiers in active service that they are not peculiar to military camps. Greater exposures give greater prominence to the diseases that result from exposure. Dietetic errors arising from many causes and combining with other influences which occasion intestinal congestions give a notable prevalence, persistence, and gravity to diarrheal diseases. Contamination of the soil and surroundings of camp, including its water supply, promotes the generation and propagation of the typhoid germ among young men who are susceptible to its influence; and the close contact of adjoining squads and companies affords the best facilities for the transmission of contagious diseases. But none of this increased prevalence and gravity is directly and unpreventably dependent on the aggregation of so many men in camps. The typhus miasm or contagion, which intensifies the danger of every other affection, and is in itself, when fully endowed with its virulence, a more deadly enemy than all the others to which the camp is subject, is not inherent in the system of military camping, but is a development from local overcrowding in individual tents, huts, barracks, or hospitals.

87. The infantry line of battle camp affords about 4.5 square yards of ground as the site of the quarters of each soldier; and when each man bivouacks, wrapped in his blanket, on this area, there is ample space for free ventilation; but if the men be grouped in squads under the shelter of tents or huts, a most unhealthy condition may exist within. A large tent is a dangerous shelter; for assuredly

as many men will be packed into it as it can hold. If fifteen or twenty men lie shoulder to shoulder during the night rebreathing a deoxygenated air laden with unwholesome exhalations from the lungs and skin, disease will in time make its appearance in the squad, no matter how thoroughly the streets and intervals of the camp may be ventilated, for the evil is within and not without the shelter canvas. The larger the squad and the smaller the superficies allowed it under shelter, the greater will be the danger of the speedy

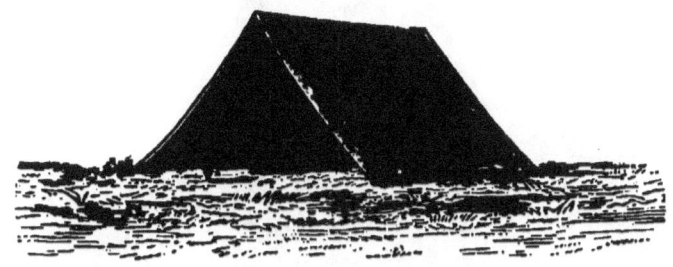

The shelter-tent.

generation of typhus or crowd-poisoning among the men. The shelter-tent carried by troops in a campaign has this great advantage over more ostentatious quarters, that it breaks up the company into small squads and scatters the men over the company area.

88. The shelter-tent is made of cotton duck, weighing eight ounces to the linear yard. Two pieces, each about 5 feet 6 inches square, are required to construct a tent. Each piece has buttons and button-holes which permit of its being fastened to any other piece. The lower edge is furnished with a loop at each corner and one at the foot of the central seam by which the piece may be pegged to the ground. Pegs and uprights are issued as part of the tent, but soldiers generally rely on the camp-ground to provide them

with extemporized substitutes. Each piece weighs 2 pounds 6 ounces; and as usually carried it is rolled with the blanket into a long cylinder, which is slung from the shoulder to the opposite hip, where the ends are tied together with the guy-rope. Two pieces, when pitched on uprights 45 or 50 inches in height, give a spread at the base of $6\frac{1}{2}$ to $7\frac{1}{2}$ feet and a covered area of 17 to 20 square feet for each of the two men.

The wedge-tent.

89. The *common, wedge-*, or *A*-tent, sometimes issued, has a spread of 8 feet 4 inches at the base, and a height and length of 6 feet 10 inches. The entrance is a perpendicular cut to the bottom in the centre of its front, which admits of each half being thrown back to expose its interior. It has a sod-cloth along its lower edge to prevent the entrance of air below. It has no provision for ventilation, and when rendered impervious in wet weather by the swelling of its fibres, the only entrance or exit for air is between the closed lapels of the doorway. This tent affords better protection than the shelter-tent, and in mild weather makes excellent quarters for two or three men.

90. The *Sibley* tent, occasionally used in our service, is a conical tent 18 feet in diameter and 13 feet high; the *conical-wall* tent gives more room for its area; but the large number of men, fifteen or twenty, crowded into these tents renders them an undesirable shelter.

91. The shelter-tent, however, is usually all that our troops have to rely upon for protection from the inclemencies of the weather during field service. If the occupation of the camp is to last for more than one night, and especially if the site or weather be damp, the men should build bedsteads of poles and forked uprights on which to spread the hay, straw, grass, or whatever they may be able to procure as a mattress. Any further stay on the same ground should be marked by improvement in the condition of the shelters and the company and regimental areas, the character of which will depend on the available material and the influences from which protection is sought. Reliance can generally be placed on the ingenuity of a body of men to make the most of the materials at command; but their efforts must as generally be checked by intelligent supervision. In seeking shelter from that which assails the senses they are likely to expose themselves to more subtile and dangerous influences which are unfelt and unknown to them. In summer camps or those of warm climates there is little danger of harmful results; the men seek the air and only such protection from the heat and glare of the sun and occasional wind and rain storms as will not interfere with the cooling influence of free ventilation. Such camps usually consist of the shelter-canvas roofing over walls of leafy willow-work, with a canopy of brushwood erected high above the tents to afford a better shade. But in winter camps, or those of cold climates, the attempt to preserve a certain degree of warmth in the interior of

the shelters is virtually an effort at the suppression of ventilation.

92. The general opinion of army medical officers is in favor of huts for occupation during cold weather; and many of those who have written on the subject have put themselves on record as insisting on 40 square and 400 cubic feet per man, with double walls, raised floors, ridge-ventilation, and warmed air supply, all of which requirements imply the presence on the camping-ground of specially provided material and labor; but huts built *by* the troops, and huts built *for* them, are two different things. In the settlement of a large army in its winter quarters the amount of transportation required for an attempt to house it as might be desired is not always available; hence the men must rely upon their shelter-tents and such materials as are afforded by the country in the vicinity of their camping-ground.

93. In the establishment of winter camps four men usu-

Army of Potomac log hut.

ally join their shelter-pieces to form a roof over low walls, generally in our well-timbered country constructed of logs. The length of this roof is 10 feet 8 inches, and its spread 7 feet; but as the canvas has to be brought down over the outer face of the logs, the interior of the hut is lessened in proportion to the thickness of its walls. Putting these at six inches gives the hut an area in the clear of 9 feet 8 inches by 6 feet, or sufficient for two double bunks with the narrowest of passage-ways between them. But what with absentees, sick, and on furlough, and the regular details for guard and picket duty, it seldom happens that more than three men pass the night in the four-pieced hut. A broad bedstead for three men is accordingly built at one end, having a space at the other of about 3 by 6 feet as a living room, on the floor of which the occasional fourth man spreads his poncho and blankets at night. The doorway opens into this space from the street; a cupboard or shelves are placed in the angle near it, and an open fireplace in the opposite wall. The preservation of the chimney is a source of much labor and constant anxiety to the occupants of the hut, as, although sometimes built of stone or brick, it is more frequently a narrow wooden shaft, with layers of clay to prevent its timbers from catching fire; yet it is deserving of all the attention bestowed on it, as when in good working order it gives a cheerful warmth to the interior while in steady operation as an efficient means of ventilation.

94. The experience of our Civil War has shown that under certain conditions these small and rudely constructed huts may give wholesome shelter to their occupants during the inclement season with far less risk of the development of diseases due to local overcrowding, or the spread of those propagated by specific causes, than the large army tents and squad barracks of the European services. These conditions are: 1st. The site of each hut should be free from moisture. The sides of the company streets and intervals should be deeply trenched, and transverse cuts made between these, uniting them and mapping out the sites of the indi-

vidual cabins. Surface drainage from higher grounds should be intercepted and turned aside. If rain fall during the period of preparation and building, the canvas should be pitched to protect the sites; otherwise they are better exposed. 2d. The floor should be cleared of all herbage, the soil well stamped with sand and gravel, and subsequently concreted; but if the site be retentive of moisture the floor should be raised about a foot from the surface by being made of split or dressed logs closely set, or of lumber which may be raised from time to time to air the underlying surface. 3d. The shelter-canvas should be so fastened that it may be unhitched with readiness when it is desirable to sun the interior. 4th. The chimney should *draw* well, as being the only means of securing ventilation. 5th. The interior should be inspected daily to insure perfect cleanliness. 6th. The camp-ground, as a whole, should be in good condition, for a satisfactory cleanliness of the person and quarters cannot be expected if the surroundings counteract all efforts to this end.

95. When these requirements are not observed the log shanty speedily degenerates into a den of filth and disease, unfit for human habitation. The soldier in cold weather is prone to burrow, and special attention must be directed to guard against this tendency. In fact, a protest must be entered against everything which is conducive to dampness of the interior. The earth must not be banked up on the outside of the logs; the floor must not be dug out to bring its level below that of the surrounding ground, nor must a side-hill be dug into to form a part of the end or side walls of the proposed hut. When a hut is converted into a half-sunk cellar by a combination of excavation inside and banking up outside, it is impossible in damp weather to preserve a healthsome, dry interior; and irrespective of diseases due solely to humidity, as catarrhs, sore throats, rheumatism, etc., there is imminent danger of the development of noxious miasms. The heat of the hut, when well warmed by its open fireplace, will recommence changes in the organic matter of the humid soil which the external winter temperature had checked, and a localized and artificial generation and evolution of malaria may be set up, prostrating the occupants with intermittents, remittents [78], and dysentery [81]; or, if the soil be less damp, the miasm evolved may be that of typhoid fever [79]. The external cold prevents emanations from the camp site as a whole, but each hut becomes a hotbed for the generation of miasms which operate with intensity—for the energies of the occupants are devoted rather to excluding the cold than to ventilating their quarters, and dissipating or diluting their dangerous atmosphere. Moreover, as the occupants begin to feel the effects of their unwholesome dwellings they drop into an apathetic condition in which all soldierly qualities are lost. Their personal appearance and surroundings cease to interest them, and they care only to pass the time in their bunks when they are not on the detail for duty. The ignorance or carelessness of company and regimental officers which permitted the construction of the dangerous dugouts manifests itself as well during their subsequent occupation: Inspections are perfunctory, filth accumulates, and ultimately the typhus miasm gives added virulence to the pre-existing causes of disease, and raises an alarm which may fortunately put an end to the insanitary conditions that are ruining the command [85].

96. The shelter-tent is invaluable in summer, when the men live in the open air and make use of the tent only as a protection to their bedstead; but it covers too small an area for comfort when, in winter, many hours of the day

have to be spent under it as in a living-room. The best log hut which the troops can construct is limited in its area by the means of roofing it; and a small increase of area under such circumstances makes all the difference between compression and comparative comfort. Generally the wear and tear of a summer's campaign renders the shelter-tents unfit for service as a winter protection, and issues of new shelter-canvas have to be made. This being the case, it would contribute much to the health and comfort of the troops if the

Winter hut for four men—the canvas roof protected by a fly which is fastened to a rail near the eaves.

Quartermaster's Department were to supply to every squad of four men a special roofing-canvas consisting of two pieces—one, 14 by 12 feet, as a roof, and the other, somewhat larger, as a fly to protect it. These would admit of the construction of a log hut having an interior measurement of 13 by 7 feet, and giving room by its length for a double bedstead at each end, and an intervening moving space between the doorway in the front wall and the fireplace opposite. With the wall six feet high, which should be its minimum, the hut would have a capacity of 700 cubic feet, the air of which would be freely renewed by the chimney-draught and the ventilating aperture in the roofing-canvas under the protection of the fly. Theoretical hygiene may object to the area and air space of the proposed hut, but the measurements are suggested advisedly, and are based on a knowledge of the military tendency to close up and occupy unoccupied space. When a hut affords possible bed and elbow room for one more man, that man will immediately become an inmate, and the hut will no longer be a hut for four, but for five, men.

97. Any tendency to crowding the huts on each other should be strenuously opposed; the minimum interval between adjacent gables should be equal to the height of the walls, six feet, while the passage between the rear walls of adjacent rows should equal the height of the ridge, about ten feet. If the company front be too small to afford this space without undue narrowing of the street, the camp should be formed in column of divisions.

98. Besides the trenching, which is intended to give a dry site to individual huts, every effort should be made to improve the general surface of the camp. Surface depressions which form pools in rainy weather should be drained and filled up. The company streets should afford a firm and dry footing when the men turn out at roll-calls. Pathways or sidewalks along the streets to the kitchens, officers' quarters, sinks, etc., should, by trenching, grading, gravelling, planking, or other means, permit of a dryshod performance of the routine business of camp life even in unfavorable weather. The perfection of this work will depend on the permanence of the camp; but the main features of the system of drainage should be worked out at once, leaving improvements to follow as the stay is prolonged.

99. Company officers are responsible for the police of the huts, kitchens, and company areas belonging to their commands, and for the personal cleanliness of their men. They should see that the interiors are kept scrupulously clean, and that the canvas roof is removed from time to time for thorough ventilation; blankets should be aired on every available occasion.

100. Personal cleanliness in winter quarters depends considerably on the facilities provided for that purpose. Huts should be built as lavatories, with safe drainage to carry off the waste water either by surface trenching or through a covered sink. A hot-water supply can be obtained by means of a boiler and barrels of water connected by circulating pipes.

101. The regimental commander is responsible for the condition of the camp as a whole; and to enable him to sustain this responsibility captains of companies are detailed in rotation as superintendents of police, under the military title of *officer of the day*. This officer has command of all the guards and prisoners, and is responsible to his superior officer for the order and cleanliness of the camp. He makes use of the prisoners in policing the grounds; and if they are insufficient for the work, fatigue details are granted him. As every day brings a fresh officer to superintend, the system is satisfactory with efficient officers.

102. With inexperienced troops and careless or incapable hygienic government a good natural site can speedily be rendered unhealthy by contamination of the soil with organic impurities. Change of site may thus become needful in a very short time, particularly in warm or moist climates or seasons, for if police parties fail to remove the dangerous material from the camp, the camp must be removed from the dangerous material. Even in the best-governed camps the occupation of winter quarters should not be prolonged after the advent of warm weather, for when the constant traffic on the company area and the steady accumulation of refuse engendered by it are remembered, soil contamination is seen to be merely a question of time.

103. The regimental area should not only be regularly and carefully policed, but the necessity for this work should be reduced to a minimum by systematic arrangements for the disposal of all the waste or refuse matters of the camp. Moreover, the intervals between regiments should be preserved in as wholesome a condition as any other part of the grounds; the labors of regimental police parties should overlap, rather than fail to meet. Besides cleaning up the regimental area, general police details attend to the condition of the sinks, remove kitchen refuse and stable manure, repair defective trenching for surface drainage, and keep the pathways passable during snowfalls and rainy weather. All gleanings from the surface should be collected in heaps and carted to a selected dumping-ground at some distance from the camp and its water supply. Covered barrels for the reception of kitchen refuse should stand on a wooden platform for the better protection of the surface from contamination by decaying organic matters; and their contents should be carted away daily. Slaughter-house offal and the carcasses of dead horses, mules, etc., should be buried at the dumping-ground.

104. The sinks in an aggregation of regimental camps are of necessity in front of the men's and in rear of the officers' quarters; but in detached camps, where there is choice of ground, they should be placed in such a position that the prevailing winds will not carry odors over the company areas. They are usually long trenches about eight feet deep and two feet wide, with the excavated earth piled on

one side, whence a part of it can readily be thrown by the police party over the daily accumulations. On the other side a short pole is laid horizontally on forked uprights at a proper height for the convenience of the men. The whole is surrounded by a thick-set hedge of brushwood, through which admission is given by an oblique or valvular entrance. Small sinks for each company are better than three or four of large size for the regiment. When the stay in camp is prolonged beyond a day or two, the horizontal pole should be superseded by box-seats open behind so that earth can be thrown in. While in winter quarters the mouth of the trench should be completely boxed with covered seats, one side being hinged to admit of layering the daily deposits with earth. When filled within two feet of the surface, each sink should be replaced by a new one, those disused being filled up and banked over to mark their site.

105. No satisfactory provision can be made to prevent soil contamination from urinary excretion. During the day the sinks are the receptacles for a large percentage of such discharges; but in bad weather their distance leads the men to find some concealed place near their quarters, often in the intervals between the huts, and at night all parts of the area are liable to contamination. Unless officers are vigilant, certain angles about the huts will soon begin to evolve ammoniacal odors. The plan of providing night-tubs is objectionable, as they cannot be of use to all without being too near to some. The medical officer should indicate such places, if any, as may be used in addition to the sinks, and the men be held to a strict observance of camp sanitary orders.

106. The water supply should be jealously guarded by the regimental medical officers. If from a stream, care should be taken that the drainage of one camp does not contaminate the supply of another. Special points below that from which the water supply is derived should be indicated for the washing of clothes, watering of horses, etc. When wells are used, their depth and distance from the sinks should be carefully considered, as also the character and incline of the strata through which they penetrate. There is no time for chemical or bacteriological analysis to determine the quality of water supplies in the field. The water is a dangerous water when the taste or odor testifies to the presence of vegetable organic matter, or when it is known that sewage, even in small quantity, enters it. It is useless to treat a water with alum, permanganate of potash, or other purifying chemicals, because if it is of such a character as to require this treatment it should be boiled. Boiling a water for ten or fifteen minutes destroys all infectious, malarial, typhoid, dysenteric, and choleric. Soldiers should be taught to fill their canteens over-night with well-boiled weak coffee as the water supply for the next day's march. Filters are made which strain out the germs of disease from an infected water. Portable filters have been used by small commands, and are useful on certain marches and expeditions; but it is doubtful if they could be relied on in time of war to supply pure water to the troops of a large army. An effort, however, should be made to accomplish it.

107. The Regulations provide for a satisfactory condition of all camps by means of official inspections; but the army looks to the medical officer for its preservation from preventable diseases. He is the sanitary officer of the command, and must render a monthly sanitary report as called for in the service of the post hospital [16]. The medical officer is not confined to this regular sanitary report as a means of bringing his recommendations to the notice of his

immediate commander and superior authority. When any fault or error in the sanitary arrangements is detected, it should be immediately reported for the action of the regimental commander.

108. When troops are *embarked on transports* the utmost care is enjoined by the Regulations for the preservation of the health of the men, for when thus crowded together in narrow limits, with imperfect means of ventilation, the absence of healthful exercise, and probably a defective diet, a tendency to typhus and scurvy is readily developed. Officers are required to enforce cleanliness as indispensable to health. When the weather permits, bedding is brought on deck every morning for airing. The men, in hot weather, are not allowed to sleep on deck or in the sun; and they are encouraged and required to take exercise on deck, in squads, by succession, when necessary. All the troops turn out at a prescribed hour in the morning without arms or uniform, and in hot weather without shoes or stockings, when every individual is inspected as to his personal cleanliness; the same personal inspection is repeated thirty minutes before sunset. On these occasions the medical officers are required to examine the men to observe whether there be any appearance of disease.

109. This chapter is fitly concluded by the instructions of Surgeon General Sternberg, issued in view of a probable invasion of Cuba by our regular and volunteer troops:

"In time of war a great responsibility rests upon medical officers of the army, for the result of a campaign may depend upon the sanitary measures adopted or neglected by the commanding generals of armies in the field. The medical officer is responsible for proper recommendations relating to the protection of the health of troops in camp or in garrison, and it is believed that as a rule medical officers of the United States army are well informed as to the necessary measures of prophylaxis, and the serious results which infallibly follow a neglect of these measures, especially when unacclimated troops are called on for service in a tropical or semi-tropical country during the sickly season. In Cuba our army will have to contend not only with malarial fevers and the usual camp diseases—typhoid fever, diarrhea, and dysentery—but they will be more or less exposed in localities where yellow fever is endemic, and under conditions extremely favorable for the development of an epidemic among unacclimated troops. In view of this danger, the attention of medical officers and of all others responsible for the health of our troops in the field is invited to the following recommendations:

"When practicable, camps should be established on high and well-drained ground not previously occupied.

"Camps should be changed to fresh ground every ten days, or oftener.

"Sinks should be dug before a camp is occupied, or as soon after as practicable. The surface of fecal matter should be covered with fresh earth or quicklime or ashes three times a day.

"New sinks should be dug and old ones filled when contents of old ones are two feet from surface of ground.

"Every man should be punished who fails to make use of the sinks.

"All kitchen refuse should be promptly buried, and perfect sanitary police maintained.

"Troops should drink only boiled or filtered water and coffee or tea (hot or cold), except where spring water can be obtained which is pronounced to be wholesome by a medical officer.

"Every case of fever should receive prompt attention.

If albumen is found in the urine of a patient with fever, it should be considered suspicious (of yellow fever), and he should be placed in an isolated tent. The discharges of patients with fever should always be disinfected at once with a solution of carbolic acid (5 per cent.), or of chlorid of lime (6 oz. to gallon of water), or with milk of lime made from quicklime.

"Whenever a case of yellow fever occurs in camp the troops should be promptly moved to a fresh camping-ground located a mile or more from the infected camp.

"No doubt typhoid fever, camp diarrhea, and probably yellow fever are frequently communicated to soldiers in camp through the agency of flies, which swarm about fecal matter and filth of all kinds deposited upon the ground, or in shallow pits, and directly convey infectious material, attached to their feet or contained in their excreta, to the food which is exposed while being prepared at the company kitchens or while being served in the mess tent. It is for this reason that a strict sanitary police is so important. Also because the water supply may be contaminated in the same way, or by the surface drainage.

"If it can be avoided, marches should not be made in the hottest part of the day—from 10 A.M. to 5 P.M.

"When called upon for duty at night or early in the morning a cup of hot coffee should be taken.

"It is unsafe to eat heartily or drink freely when greatly fatigued or overheated. Ripe fruit may be eaten in moderation, but green or over-ripe fruit will give rise to bowel complaint. Food should be thoroughly cooked and free from fermentation or putrefactive changes.

"In decidedly malarious localities from three to five grains of quinine may be taken in the early morning as a prophylactic; but the taking of quinine as a routine practice should only be recommended under exceptional circumstances.

"Light woollen underclothing should be worn; and when a soldier's clothing or bedding becomes damp from exposure to rain or heavy dew, the first opportunity should be taken to dry it in the sun or by fires."

CHAPTER IV.

GENERAL HOSPITAL SERVICE.

110. *General hospitals* are those established at points distant from the field of actual warfare. The first of these to which the wounded man is transferred is usually that at the *base of supplies;* but his stay here is seldom of long duration, as this hospital is in reality merely a resting and shipping point on the route to places of greater security. It is organized on the general hospital system, and may indeed be viewed as the general hospital of the army to which it is attached; but as its existence in a locality is dependent on military movements, its accommodations and appointments have usually more of the character of a field than of a general hospital. Associated with it are *hospital boats* or trains of *hospital cars,* all of which are manned by assignments from the Surgeon General's office.

111. A general hospital is practically an expansion of the post hospital. The latter consists of an administration building and one or two attached wards which may be lengthened or shortened, within limits, to suit their capacity to the requirements of the time and place. The former consists of a *series* of long pavilion wards, each capable of accommodating forty to sixty patients, with an administration building, kitchens, laundry, stables, repair-shops, etc., and quarters for the officers, employees, and guard, on a scale proportioned to the size of the hospital. At the beginning of our Civil War, hotels, churches, court-houses, factories, and other large buildings were used as general hospitals; but the advantages of the pavilion system were soon recognized, and extemporized hospitals became replaced by special constructions. In these the wards were variously arranged to secure a full allowance of ventilation and sunlight for each, and at the same time keep them within convenient distance of the offices and other buildings. In some hospitals they were arranged in a line, with their gables facing the front and rear, the administration building in the centre of the line, and the other buildings disposed in the rear. In others they were placed *en échelon* in the form of a V, with the administration building at the apex, the kitchens and dining-rooms in the interior, and the other buildings closing in the base. In others, again, the pavilions enclosed a circular or oblong space, the administration building occupying a central position among the wards, and the other buildings within the enclosure. Adjacent wards were separated from each other by a clear space of about thirty feet. The wards, kitchens, dining-rooms, and offices were connected by means of a covered corridor or walk. The plan of the *Hicks Hospital*, Baltimore, Md., submitted on page 72, illustrates one method of arrangement.

112. Each ward of a general hospital was a ridge-ventilated pavilion from 145 to 187 feet long, 24 wide, and 14 to the eaves. The smaller length, for forty patients, was generally preferred to the larger one, for sixty. At each end of the ward there were partitioned off two small rooms 9 by 11 feet, with a six-foot passage-way. Those at the attached end were used as a wardmaster's room and pantry; those at the free end as a bathroom and water-closet. In some instances the latter were cut off from the ward by a passage-way giving cross ventilation, and in others they were attached to the lateral aspect of the pavilion at one of the angles of its free end. The floors were raised at least 18

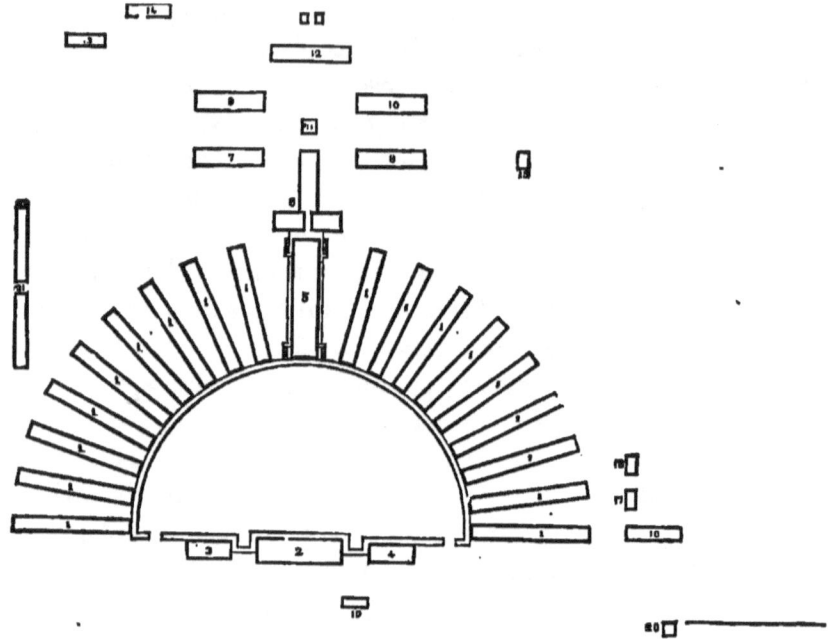

Ground Plan of Hicks Hospital, Baltimore, Md.—1, 1, 1, 1, wards; 2, administration building; 3, linen-room; 4, dispensary and operating-room; 5, dining-hall; 6, kitchen and laundry; 7, ward for detailed men; 8, knapsack-room; 9, subsistence storehouse; 10, quartermaster's storehouse; 11, tank; 12, quarters for the guard; 13, stable; 14, wagon-house; 15, sutler's store; 16, steward's quarters; 17, 18, officers' quarters (of which there are several not shown on the plan); 19, guard-room; 20, guard-house near entrance gate; 21, workshop; 22, contagion-ward,—this was more distant than is represented. The wards, dining-room, and administration building are connected by a covered way.

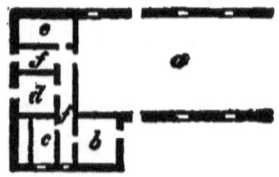

Water-closet attached to lateral aspect of free end of a ward: *a*, interior of ward; *b*, water-closet; *c*, lavatory and bathroom; *d*, pantry; *e*, wardmaster's room; *f, f*, ventilating-hall and passage-ways.

inches from the ground, and had free ventilation underneath. The beds were placed at regular intervals from each other, two occupying the floor space between adjacent windows. During warm and mild weather the wards were ventilated by the ridge. The opening, about one and a half feet wide, extended the whole length of the building, and was protected by a ridge-roof which lapped well over it on either side. During winter the ridge was closed and ventilation by shafts and special fresh-air inlets was substituted. The

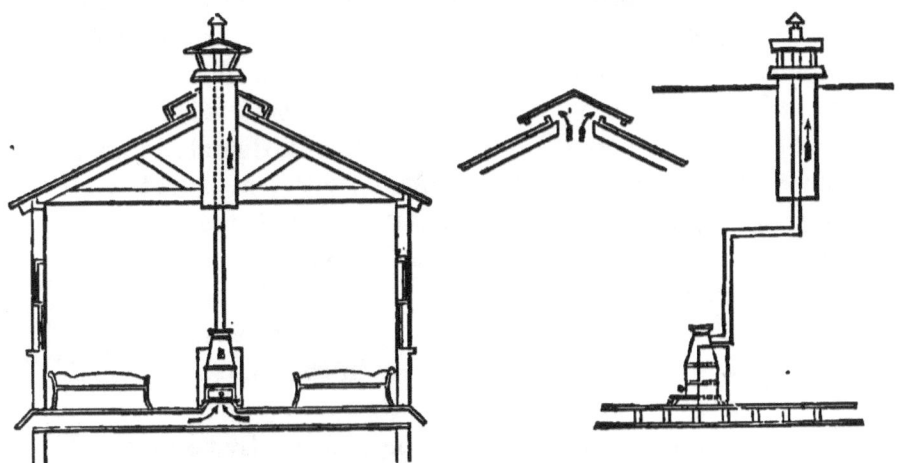

Ventilation and heating of a ridge-ventilated ward.

inlets were boxed channels from the side walls opening beneath the stoves, which were each partially surrounded by a jacket of sheet-iron or zinc. The air, more or less warmed in its passage into the ward, became diffused and was ultimately drafted through ventilating-shafts 18 inches square which extended from the level of the tie-beams to beyond the ridge. The heat of the stove-pipe was utilized in promoting the escape of foul air through these shafts.

113. The size of a ward, as compared with the number

of occupied beds in it, is a matter of importance. Many of the diseases which in former days increased the mortality in hospitals were due to overcrowding. Erysipelas has become infrequent, and hospital gangrene unknown [372], since a proper amount of space has been assigned to each bed. Typhus fever has also become a disease of the past, as well as that typhus-like character which overcrowding impressed on pneumonia and all other febrile diseases [85]. The air space in the hospitals of the war was from 924 to 1,033 feet per bed; but all the beds were rarely occupied at the same time. More space is required for suppurating wounds, infectious diseases, and such as confine the patient to bed than for trivial cases or convalescents to whom the ward is merely a sleeping-room. When a liberal air space is afforded, ventilation can be effected with less risk of creating a draught. If a ward give 3,000 feet of space [230] to each patient, its air would have to be renewed only once per hour to insure good ventilation; whereas, if it be crowded with one man for every 500 feet of its capacity, the air would have to be renewed six times in the course of an hour to preserve its quality.

114. The administration building was usually two-storied, and contained the general office, office of the surgeon in charge, chaplain's office, dispensary, linen and store rooms, lodging-rooms for officers, etc. The kitchen was divided into two parts, the larger for the preparation of ordinary diet, the smaller for extra diet. The dining-room seated a number equal to two-thirds of the number of beds; it communicated with the kitchen usually by the centre of one of its sides. The subsistence and quartermaster's store-room contained boxes and shelves for the various parts of the ration, a room for clothing, and on its second story lodging-rooms for the cooks; an ice-house was connected with it.

A knapsack-house received the effects of the patients while in hospital. The laundry was a two-storied building, having quarters for the laundresses on the second floor. Special quarters, including dining-room and kitchen, were provided for female nurses. The other buildings consisted of quarters for officers; an operating-room and a dead-house, both lighted by skylights, the former near the administration building, the latter in a retired part of the grounds; a chapel, with library and reading-room attached; guards' quarters, stables, repair-shops, etc.

115. When the water supply was not derived from the mains of a city, steam was usually employed to raise it from the wells, springs, or streams which formed its source, and in this case the engine was generally situated near the kitchen and laundry that the steam might be made available in cooking and the power utilized in working the washing and mangling machines. It was usually considered advisable to have some reserve tanks or cisterns kept always full in case of danger from fire.

116. Rain-water is wholesome water if properly collected and stored. The roofs or other shedding surfaces should be clean; if they are foul, the first fall of a shower should be run to waste by a cut-off, if a sedimenting cistern or filter be not interposed between the watershed and the reservoir. *Cisterns* are usually constructed of brick, lined with Portland cement, or of wood, generally cypress wood. Unless care is exercised in excluding the washings of the watershed, the cistern will soon accumulate a thick sediment of foul mud, which must be cleaned out from time to time. In warm weather, when the water-level in the cistern is low, this sediment may seriously affect the quality of the water. Underground cisterns, from their cooler situation, are less prone to suffer from the fermentation of an accumu-

lated sediment. Moreover, the mineral or earthy matters of which the underground cistern is constructed introduce into the stored water certain bacteria which transmute ammonia into nitric acid. These are called the micro-organisms or *bacteria of nitrification*. Organic matter that may be present in the water from the air-washing which it has effected, or from foul accumulations on the watershed, in the conductors, or in the cistern itself, is decomposed into ammonia, and this is subsequently transformed into nitric acid. The tendency of the water during its storage in the cistern is to improvement; but it is important to observe that this does not hold good in wooden tanks, unless the bacteria of nitrification are introduced, as by throwing into the cistern a quantity of clean gravel.

117. *Surface water*, as from *rivers, lakes, ponds*, etc., is often impure from filth washed from the watershed. The *subsoil water*, tapped by *shallow wells*, is free from the turbidity often found in surface waters; but it is not generally accepted as a wholesome water. If the soil be a clean sand in an unsettled locality the water may be as good as any filtered cistern water; but if it be impure from the soakage into it of the wastes of human life and occupation, the water will be more or less tainted with these impurities. The water of *deep wells* is usually organically pure, but often so charged with saline matters as to be undesirable as a potable supply.

118. There is no easily performed chemical or other test for organic matter in a water. If a quart bottle half filled with the water at a temperature of 70° or 80° Fahr. be shaken vigorously for a few minutes and then placed to the nose, an organic *odor* may be detected in the air of the bottle if the water is of doubtful or bad quality; but bad waters do not always have an odor. The best of the easy modes

of chemical inquiry is to *burn the residue*. Evaporate 100 c.c. of the water to dryness in a platinum or porcelain capsule; then ignite the dish over the flame of a lamp. If there be no blackening, or at most only a darkening of the residue, which is speedily dissipated by a continuance of the heat, the water is probably good. If the thin crust of the residue blacken all over, and the carbon be afterward dissipated with difficulty, the water has probably an excess of vegetable matter. If, in addition to the blackening, nitrous fumes are evolved, and the carbon sparkles in points with the energy of its combustion, the water may be suspected of containing organic matter of animal origin. The organic matter found in drinking-water may be of a harmless or dangerous character; but it must be conceded that where there is much organic matter the likelihood of the presence of dangerous matter is greater than where there is little.

119. The presence of salts of lime and magnesia gives a water the quality of *hardness*. Soap does not form a lather with hard water until the lime and magnesia have been precipitated in the form of curdy salts of the fatty acids of the soap. When a hard water is boiled, white flakes of carbonate of lime appear in it, and its *temporary* hardness is removed. When the earthy salts are present in the water as sulphates the hardness caused by them is called *permanent* because it is not removed by boiling. Such waters induce diarrhea, particularly in those who are unaccustomed to their use. *Soft waters* contain but little of these earthy salts.

120. As water is frequently distributed by leaden pipes, and sometimes stored in lead-lined cisterns, the possibility of the solution of poisonous quantities of the metal must be held in view. The symptoms are violent neuralgic pains in the abdomen, simulating *colic*, but oftentimes affecting

also the limbs and trunk, with constipation and gradual loss of strength. When lead is used for service-pipes the water which has stood in the pipes over-night should be run to waste before drawing a supply for use. Rain and other soft waters act on lead with facility. When metal is used for cisterns, iron, coated with asphalt paint or black varnish, should be employed. Zinc, which forms the protective coating of the iron in galvanized tanks and pipes, is acted on by most waters, but without producing notable harmful effects on the consumers.

121. When the water supply was adequate it was introduced into the water-closets attached to the wards and into the latrines for the use of convalescents and others. Water-tight boxes, which were emptied and cleaned regularly, were used in the absence of a water service. The *earth-closet*, consisting of a closet-seat over a pail or other small portable receptacle, with dry earth as a deodorant, was not brought into general use until a few years after the war. The drains and sewers of hospitals within municipal bounds were connected with the general sewerage system. In other cases the sewers of the hospitals found an outlet into some neighboring stream or tide-water; but where no satisfactory outlet was obtainable, the sewer terminated in a cess-pool from which liquids percolated or overflowed by a suitable conduit into a natural incline leading from the hospital, and solids were removed from time to time as they accumulated.

122. The sewerage system includes *water-closet basins*, each with a *water-seal* to prevent the inflow of foul air through their discharge-pipes, a *soil-pipe* leading downward from the water-closets on the various floors and receiving the waste-pipes from bath-tubs, kitchen traps, and other water fixtures, and a *drain* connecting the lower end of the soil-pipe with the sewer.

123. *Water-closets* are of several forms and many varieties. The *hopper-closet* is the simplest, because it has no mechanical parts to get out of order. It consists of a funnel-shaped bowl which leads the deposits into the water of an S-shaped trap. Its efficiency depends on the quantity of water available and the manner of its distribution from the flushing rim over the curved sides of the basin. *Pan-closets* and *valve-closets* are objectionable as liable to become

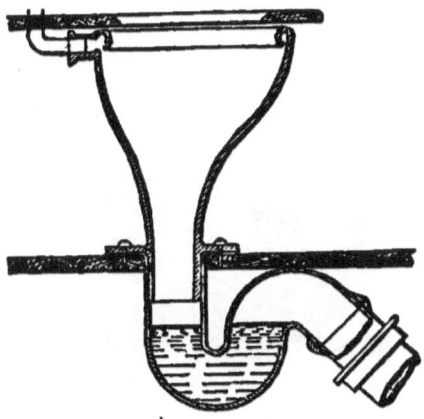

Hopper-closet with water-seal.

foul and out of order. In the *plunger-closet* the outlet from the bowl is at the side instead of below, and is closed by a heavy metal piston or cylinder which, on being raised, discharges the contents into an S-shaped trap leading into the soil-pipe. Leakage from the bowl sometimes occurs when the plunger, on account of fouling, fails to completely close the outlet.

124. Bath-tubs, wash-basins, and the fixtures of the laundry are connected with the soil-pipe usually by lead pipes one to two inches in diameter, trapped by a deep S-shaped bend close to the aperture of outflow. Small overflow pipes are generally provided, and these are either trapped themselves or connected with the main outflow on the near side of the bend. Kitchen and pantry sinks have two-inch outflows provided with a strainer and trapped close to the bottom of the sink. As the trap becomes sometimes choked with sediment and accumulations of grease, it has usually a screwed cap on its convexity by which it may be entered and cleaned. Hot water often carries liquefied grease to a considerable distance along the outflow before it becomes congealed. An occasional flushing with a solution of soda

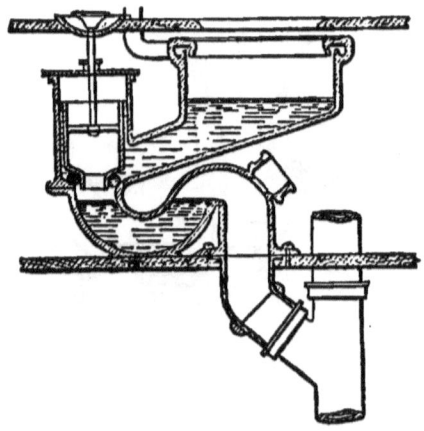

Plunger-closet with water in basin and in S-trap and hub on soil-pipe side of trap for attachment of vent-pipe.

or potash will tend to clear the two-inch pipe, and the ammoniacal fermentation of water-closet discharges has a similar scouring influence on the main drain.

125. *Soil-pipes* descend vertically through the building from above the roof to the cellar. Each is open at the top and ends below in a curve which connects it with the house-drain. Water-closet outflows and the wastes of bath-tubs, wash-basins, and kitchen trays, etc., join it by Y-shaped branches. All junctions must be solidly made. Iron pipes are joined by pouring melted lead into the sockets when the lengths are in position, a small packing of oakum having been previously introduced to prevent the liquid metal from penetrating into the interior; and when cold the lead is driven securely home by a hammer and caulking-iron. Lead is joined to iron by tipping the leaden pipe with a brass ferrule, which is afterward caulked into the iron with melted lead.

126. It is better to have these pipes in sight than boarded up, as any flaw in the plumbing is more readily detected. When a leak is suspected the peppermint test is recommended for its discovery. A fluid ounce of this volatile oil, or a corresponding quantity of its essence, is poured into the upper end of the soil-pipe, and a flush of water is sent down after it. The odor of the oil is so penetrating that it speedily makes itself felt at any leaky point or flaw in the system of pipes; but the search for its presence must be conducted by one who has kept himself free from any recent contact with the oil, and the man who made use of the test-liquid must remain at his post until the end of the investigation, lest he carry an odor with him which would interfere with the discovery of a leak.

127. The *house-drain*, of iron pipe six inches in diameter, should traverse the building along the ceiling or walls of the cellar, or, if it be needful to place it under the floor, it should be laid in a concreted trench with a fall of at least 1 in 100, and a cover which can easily be removed for inspection. Outside the walls the drain may be either of iron or vitrified pipe. Iron should be used if the ground is liable to sag, or if the drain passes within the drainage area of a well-water supply, or near the roots of trees, which, in their search for water, will penetrate the joints of vitri-

fied pipes and choke their interior. The term *drain*, which custom has applied to this pipe, is sometimes misleading. *Drainage* is the system by which the surface and subsoil are relieved from an excess of moisture, and drains are properly the tile-pipe or other channels by which this is effected; while *sewers* are the channels of the *sewerage* system by which *sewage* is removed. Evidently the pipe in question is rather a house-sewer than a house-drain.

128. At some convenient point, either inside or outside the walls, this house-drain or sewer should be trapped by a deep-sealed S-bend to cut off all communication between the air of the common sewers and that of the system of pipes within the building. This trap should be well protected against freezing in cold weather. When *rain-conductors* join the main drain on the near side of its trap they require no special trapping, but if their junction be effected on the far side they will, if untrapped, become ventilators for the sewers, and may diffuse unwholesome gases throughout the upper part of the building.

129. The water-closet system presents two guards against the entrance of sewer air into a building—the main trap on the house-drain, and the traps on the individual waste-pipes. Should the former be forced by some sudden air pressure in the common sewers, the foul air enters the soil-pipe, but on account of its open upper end no pressure is brought to bear on the interior traps. But the interior of the pipes on the house side of the trap on the main drain may be so coated with fermenting organic matters that the air contained in them may differ but little in quality from that of the sewers. The open end of the soil-pipe above the roof has of itself no ventilating or purifying influence. The evaporation of the water in the trap of an unused bath-tub or wash-bowl might therefore give entrance to very unwholesome gases from the soil-pipe. It is advisable on this account to have the whole system of pipes on the hither side of the main trap as freely ventilated as possible. This is accomplished by means of a *fresh-air inlet* into the drain. The inlet usually takes the form of a four-inch iron pipe which extends from some distance above the surface of the ground to the drain, tapping the latter at a convenient point between the lower end of the soil-pipe and the main trap. Its free end is covered with a cowl or raised cap to prevent the entrance of foreign matter. The warmth of the soil-pipe in the interior of the house and the aspiratory force of the wind on the open mouth of its upper end above the roof establish a constant current of fresh air through it which materially lessens the danger attaching to accidentally unsealed traps. Unsealing is sure to occur if the plumbing fixture remains unused for a certain length of time, depending on the warmth of the weather or room, and the depth of the water in the trap. The remedy in this case is obvious.

130. Sometimes, however, water-traps are unsealed by what is called *siphonage*. When the upper bend of an S-trap becomes filled, full bore, by a sudden discharge of water, the trap acts as a siphon, and may draw off so much of the water as to leave itself with its seal broken. As might be expected, small pipes and shallow seals are more likely to be siphoned than large pipes and deep seals. Again, the sudden rush of a discharge from a water-closet through the soil-pipe may suck out the water-seal of a neighboring trap. Both of these accidents are prevented by means of a vent-pipe on the soil-pipe side of the trap, which permits of the entrance of air in the one case to break the siphon, and in the other to fill the vacuum caused by suction. Vent-pipes from traps unite into a single pipe, which may end above the roof like the soil-pipe, or open into the latter at a point above the highest fixture. Mechanical devices are sometimes used to increase the efficiency of the water-seal or guard against its loss. Thus, in Bower's trap, a rubber ball is buoyed by the water against the mouth of the pipe leading from the fixture. The discharge, in passing, temporarily displaces it, but it immediately resumes its guard, closing the aperture so long as enough of water remains in the trap to float it into position.

 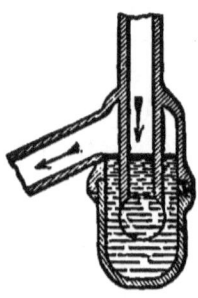

S-trap—showing water-seal, screw cap on convexity for convenience in cleaning, and vent-pipe to prevent siphoning.

Water-seal with rubber ball acting as a valve. Bower's trap.

131. *Latrines* consist of a series of hopper or other closets over a brick trench or iron receptacle containing water. From time to time the plug which guards the outlet of the receptacle is raised and the contents are flushed out through an S-trap into the soil-pipe.

132. Sewers are generally built of considerable size, to carry off the rainfall as well as the sewage; and heavy rains do much good from time to time in flushing out and cleansing their channels. The rain-leaders from a building usually enter the house-drain on the near side of the trap [128]; but in climates where the winters are not severe they may end in surface channels which carry the water through a grating into a catch-box for gravel, the overflow passing through a trapped drain to the sewer.

133. Ground areas on which rain-water would otherwise accumulate are often drained by means of what is called the *bell-trap*. A perforated metal plate permits the water to enter a basin which lies underneath it, and when the water rises to a certain height in the basin it overflows into

The bell-trap.

a central pipe which carries it into the house drain. A hemispherical cup attached by its bottom to the under surface of the plate makes a loosely fitting cover for the mouth of the pipe, and by dipping into the water contained in the basin prevents the escape of emanations from the drain. It should be remembered that traps of this kind are readily unsealed by evaporation.

134. The surgeon in charge of a general hospital has full and complete military command over the persons and property connected with it. At small hospitals he is his own *executive officer;* but in large establishments an officer is detailed to aid him in his supervision. The duties of this officer are those of an adjutant to a commanding officer, with those of subsistence officer and quartermaster superadded. He has charge of the office and records, and of the men detailed as clerks and orderlies; he keeps the clothing and other accounts of the detachment and of the detached men; he supervises the preparation of all regular reports, promulgates all orders, and conducts the general correspondence. He distributes the patients received for admission, and looks after the general well-being of the establishment as aid to his superior. The same reports and returns are rendered and books of record kept as at a post hospital.

135. Several *stewards*, with clerical assistance, are required in the service of a large general hospital. One takes charge of the books and papers relating to the military government of the establishment,—he is practically the sergeant-major of the command; a second attends to matters of subsistence; a third to quartermaster's property and the issue of clothing; and a fourth to medical and hospital property and supplies. A steward superintends the work of the dispensary; one has general charge of the operating-room, wards, and dead-house; one looks after the kitchen and dining-room; and one attends to the laundry and the work of disinfection.

136. Each *ward surgeon* is responsible for the professional treatment and general comfort of his patients; for the police of his ward, the care of its property, and the faithful discharge of their duties by his subordinates. He makes a record of all cases of professional interest, and sends a morning report to the executive officer, stating all changes and recommending others, such as the return to duty, furlough, discharge, or transfer of particular individuals. An *officer of the day* is detailed daily by roster from the number of the ward surgeons. This officer must not be absent from the hospital during his tour of duty. He admits patients in the absence of the executive officer, and prescribes in cases of emergency in the absence of the ward surgeons. He inspects the meals and visits the wards at bedtime, and again after midnight, to regulate lights and note the vigilance of ward attendants. The detail for guard

is under his command to enable him to enforce discipline at all times; but if the guard consists of a special body of troops, its senior officer is held responsible for the general police of the grounds and the preservation of order within the limits of the command.

137. The *chaplain*, in addition to duties of a purely spiritual character, usually keeps a record of special patients, with the post-office addresses of the nearest relatives; he superintends the postal service, library, reading-room, and cemetery.

138. Each ward is under the care of an acting steward or private assigned as *wardmaster*, who is responsible for the comfort, diet, and medication of the patients, the performance of their duty by the nurses, the preservation of the ward property, the regulation of the fires, lights, and ventilation, and the cleanliness of the bed-linen and clothing, lavatory, bath, water-closets, etc. Two nurses are sufficient for a pavilion of fifty beds when the cases are not of an acute character; but three, four, five, or more may be required, according to circumstances. These are detailed from the allowance provided for the hospital by the Surgeon General.

139. The *hospital fund* of the general hospital differs in no respect in its management from that of the post [14]. Where hospital gardens are cultivated this fund is usually capable of supplying all the needs of the extra-diet kitchen. One of the most important duties of the hospital steward is to see that the hospital fund does not suffer from ignorance or want of economy in the kitchen.

140. Special care is needful at a large general hospital to guard against danger from fire. Every member of the hospital corps should at all times be on the alert for the general protection. Full buckets and axes should be kept in

each building, with a suitable length of rubber hose for attachment to the water service. These provisions suffice for the suppression of fire when discovered in its incipiency; but to provide for the protection of patients and property on an occasion of general danger, the whole command should be organized and drilled, from time to time, as a fire brigade.

PART II.

ANATOMY AND PHYSIOLOGY.

141. The body consists of a multiplicity of living tissues aggregated into organs, each of which has its special function to perform in order to preserve the integrity of the whole. These organs may be divided into three sets:

I. Those of *locomotion* consist of the bony skeleton, which gives form and stability; the joints, which permit of motion between the bones, and the masses of contractile flesh or muscle, which effect the motion.

II. Those concerned in the processes of *organic life* consist of an alimentary system, which renews the blood by elaborating it from the raw material of food; a nutritive apparatus, which feeds the various parts by the circulation of a liquid, the blood, and a depurative or excretory system, consisting of lungs, skin, kidneys, etc., which removes from the blood the impurities gathered in its course.

III. Those of the *administrative system* include the organs of the senses and the nervous system, from which emanate all the powers of vitality.

CHAPTER I.

THE LOCOMOTOR SYSTEM.

142. *Bone* consists of animal tissue permeated with earthy salts, chiefly phosphate of lime. The animal tissue may be demonstrated by dissolving the earthy salts in dilute hydrochloric acid. Each bone is covered with a strong fibrous membrane, the *periosteum*, in which the blood-vessels of the bone subdivide [183] before entering the bony tissues.

143. The *muscles* constitute the flesh or *lean* of the animal tissues. Each consists of a mass of parallel fibres aggregated into bundles and bound together by a fine elastic webbing which is called the *areolar, cellular,* or *connective* tissue. Every fibre recognized by the eye is composed of a vast number of microscopic fibrils, each of which is marked with close-set, transverse lines; and when the fibril con-

Bundles of striated muscular fibrils.

tracts the lines come nearer to each other, as the coils of a spring are closer when they are compressed than when they are expanded. These markings are called *striæ;* and the muscles that present them, *striated* muscles. All the *voluntary* muscles, or those under the control of the will, are striated. Muscular fibres which contract independently of the power of the will, such as those which move the intestinal contents, are flattened, band-like fibres without striæ.

144. At each end of a voluntary muscle the contractile fibres become blended with strong fibrous tissue, which interweaves with the periosteum of the bone and gives the muscular fibres a strong attachment. In some instances, as in the muscles of the forearm, the fibrous tissue assumes the rounded form of a tendon or sinew. Generally one of the two attachments of a muscle is more readily moved than the other. A muscle which extends from the shoulder to the forearm will, by its contraction, bend the elbow, and raise the hand to the shoulder; but if the hand be made the fixed point, as when we seize a bough or bar overhead, and endeavor to raise the body by sheer strength of arm, the same muscle, by its contraction, will raise the shoulder to the hand.

145. Although the contractility of a muscle is ordinarily exhibited only through the influence of the will, the tendency to contraction is continually in force. When the belly of a muscle is cut across, the fibres contract toward their point of attachment, and a gaping wound is the result. When a bone, as the arm-bone, is fractured, the muscles which extend from above to below the fracture may, by their contraction, cause the broken ends to override and give rise to shortening of the limb.

146. The *back-bone* or *vertebral column* extends from the skull down along the middle line of the back. If the fingers be drawn along this line a number of bony prominences, called *spinous processes*, will be felt, each of which belongs to one of the bones composing the column. The motion between adjoining bones is slight, but the combined motion of the whole is considerable. In position in the body these bones or vertebræ constitute a pliant pillar about twenty-seven inches long, rounded and smooth in front, irregular from many projections behind and at the sides, and having in its interior a long canal formed by the apposition of circular apertures in the individual bones. There are seven vertebræ in the neck, or *cervical* region; twelve in the back, or *dorsal* region; and five in the loins, or *lumbar* region. They are held in position in the erect attitude, and moved as required by powerful muscles inserted into their spinous and other processes; and these muscles afford protection from injury by acting as a padding to the column. Suspended in the canal of the column is the *spinal cord*, or that portion of the nervous system from which are given off most of the nerves that superintend motion and transmit sensation. Apertures are left between the vertebræ along each side of the column for the passage of nerves from the cord to the various organs and tissues.

147. This flexible pillar is supported upon a bone called the *sacrum*, which is wedged into a triangular interval between the hip-bones behind. It contains within its canal the final breaking up of the spinal cord for the nerve supply of the lower extremities. The sacrum is tipped below by a small bone, the *coccyx*. See page 102.

148. The *skull* is divided into the cranium and the face. The bony plates forming the vault of the cranium consist of two layers of compact tissue and a thin interlying layer of spongy bone. The cranial bones are united by close-fitting sutures, a series of projections and notches on one bone fitting into a corresponding series on the adjoining bone. The brain or organ of the intelligence and nervous power is contained in the cranium, and is continuous below with the spinal cord through a large circular opening in the base immediately over the canal of the vertebral column.

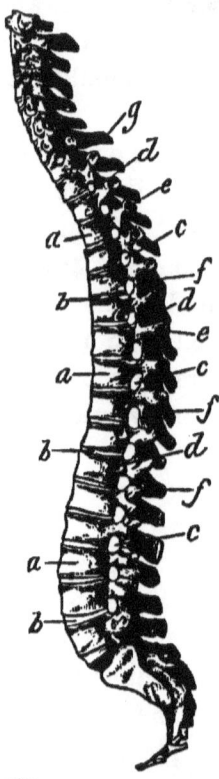

Side view of vertebral column: *a*, bodies of the vertebræ; *b*, cartilages between vertebræ; *c*, apertures for nerves; *d*, facets for ends of ribs; *e*, facets for support of ribs; *f*, spinous processes projecting behind; *g*, prominent spine of seventh cervical vertebra.

149. The bones of the face are very irregular in form, and, with the exception of the lower jaw, are closely sutured together. The rounded head of the *lower jaw* may be felt in front of the lobe of the ear, and when the mouth is opened the finger may be pressed into the back part of the cavity of the joint [476].

150. The cervical vertebræ are deeply embedded in the muscles by which the movements of the head are effected. These muscles surround and protect the *larynx*, or organ of voice, the cartilages of which project in the middle line in front [224]; the *trachea*, or windpipe, which stretches from the lower part of the larynx down behind the notch in the upper part of the breast-bone [222]; the *œsophagus*, or gullet, which lies behind the larynx and trachea, and the large blood-vessels and nerves which are embedded on either side of these tubes. One of the most noteworthy of the muscles of the neck is the *sterno-mastoid*, which stretches as a firm, fleshy mass from the junction of the breast-bone and collar-bone on either side upward and backward to the bony prominence behind the ear, its outline being distinctly marked on the surface when the head is turned to the opposite side.

151. The dorsal vertebræ have connected with them the bones which form the framework of the chest. These consist of twelve *ribs* on each side and the *sternum*, or breast-bone, in front. The outline of the individual ribs and their arrangement as a whole can usually be made out without difficulty on the person. The seven upper ribs on each side are attached in front to the margin of the sternum by means of a cartilaginous prolongation of the bony tissue. Cartilage is an opaque, bluish-white elastic substance, familiarly known as gristle. The cartilages of the eighth, ninth, and tenth ribs curve upward on either side of the *epigastrium*, or

pit of the stomach, and unite with that of the seventh rib. The eleventh and twelfth are called floating ribs, because they have no fixed attachment in front. The cartilages add greatly to the elasticity of the ribs, lessening the risk of fracture and injury to the contents of the chest. The movements of the ribs in inspiration are upward and outward, enlarging the capacity of the chest in all directions. The chest contains the lungs, heart, and great blood-vessels, and the gullet on its way downward to the stomach.

152. The *clavicle* or collar-bone lies between the upper part of the breast-bone and the point of the shoulder, where it is united to a projection of the shoulder-blade, called the *acromion* process. The triangular outline of the flattened *scapula* or shoulder-blade can be defined by the eyes or by the pressure of the fingers, its base forming part of the breadth of the shoulder, and its apex reaching a little below the eighth rib. Immediately below the junction of the collar-bone and shoulder-blade at the point of the shoulder, the bony tissue of the latter forms a shallow depression, the *glenoid cavity* [478], in which the rounded head of the arm-bone has free play for its movements.

153. The bony surfaces which move on each other, constituting a *joint*, are bound together by strong fibrous tissue which forms a capsule around them to prevent dislocation while admitting of the needful degree of motion. Where the strain is greatest the fibrous tissue is strengthened by interlaced bands which are called *ligaments*. The joint-ends of the bones are in the living body coated with a layer of elastic cartilage, and this is covered over with a thin, smooth membrane which gives a highly polished surface to the interior of the joint and secretes a lubricating liquid called *synovia*. The interior of the capsule has a similar lining, while externally it is strengthened by the apposition of neighboring muscles. The capsule of the shoulder-joint is strongly supported by muscles: in front, the *pectoral* muscles, which converge from the front of the chest to be inserted into the inner side of the arm-bone below the capsule; behind, the muscles which converge from the *scapular* region to be inserted into the upper and back part of the arm-bone; and on the outer side forming a cap to the joint, the *deltoid* muscle, which curves from the clavicle and scapula above the joint to the outer side of the arm-bone a little above its middle. Nevertheless the shallowness of the glenoid cavity and the laxness of the capsule which give to this joint its freedom of motion, render it correspondingly liable to dislocation [478].

154. The muscles mentioned in the last paragraph have an interest in connection with fractures of the arm. The deltoid raises the limb, pulling it away from the side; the pectoral muscles act in the opposite direction; the scapular muscles raise it. When the arm is broken just below the line of the arm-pit, the front wall of which is formed by the pectoral muscles in their passage from the chest to the arm, these muscles *drag the upper fragment inward*, while the deltoid draws the lower fragment, to which it is attached, upward and outward. On the other hand, if the fracture is above the attachment of the pectoral muscles, the *upper fragment will be displaced outward* by the power of the scapular muscles.

155. The *humerus* or bone of the arm consists of a cylindrical shaft expanded at its ends to enter into the formation of the joints. The shaft, like that of all the long bones, consists of a compact bony tissue hollowed along the centre into a canal containing a fatty substance or marrow. The ends are of spongy tissue covered with harder bone. The upper end or head [478] has a rounded surface on its inner aspect for articulation with the glenoid cavity, and rough prominences or *tuberosities* on its outer aspect for the attachment of muscles. Toward its lower end the shaft becomes flattened from before backward, to have greater breadth for its hinge-like joint with the bones of the forearm; and on either side of the articular surface is a projection or *condyle* for the attachment of muscles. Both condyles can be outlined by the fingers, but the inner is more prominent than the outer.

156. The bones of the forearm are the *radius* on the outer or thumb side, and the *ulna* on the inner or little-finger side. The latter, at its upper end, is scooped out from before backward into a semicircular surface which hinges with the lower end of the humerus. When the forearm is bent on the arm the back part of this semicircular notch forms the *olecranon* process, or point of the elbow. The upper end of the radius is small and rounded, but it may be felt below the external condyle on the posterior and outer aspect of the joint. The muscles on the back of the arm *extend* or straighten the forearm at the elbow. The muscles in front *flex* or bend it. One of these flexors, the *biceps*, passes to the radius just below the elbow-joint; it forms the fleshy mass on the front of the arm. Other flexors stretch from the inner condyle along the front of the forearm to operate on the fingers; these become tendinous as they approach the wrist.

157. In addition to accompanying the ulna in its hinge-motion on the humerus, the radius has a kind of circular motion on the ulna by which the hand is rotated. When certain of the muscles on the front of the forearm contract, the lower part of the radius is rolled over and in front of the ulna, so that the *back* of the hand looks *upward or to the front;* this is called *pronation*. When certain of the posterior muscles contract, the motion is reversed, throwing the *palm upward or to the front;* this is called *supination.* When the fingers of the observer are placed firmly on the head of the radius, while the hand of the patient is alternately pronated and supinated, the rolling motion of the bone is easily observed [460].

158. Eight small bones, fitted closely together in two rows, form the wrist or *carpus*. The upper row articulates with the lower end of the radius; the lower row with the bases of the five metacarpal bones. All these joints are encrusted with cartilage and lined with synovial membranes. The *first metacarpal* bone is the uppermost of the three bones which enter into the formation of the thumb. The four other metacarpal bones form the framework of the palm of the hand. At the knuckles they articulate with the bones of the fingers. The four fingers consist each of three pieces or *phalanges* jointed together with hinge-joints; the thumb has but two phalanges.

Back view of right wrist and hand: *a*, radius; *b*, ulna; *c*, bones of carpus; *d*, ligaments, shown only on left side of illustration; *e, e,* row of metacarpal bones; *f, g, h,* 1st, 2d, and 3d phalanges.

159. The tendons of the flexor muscles on leaving the front of the forearm pass under a strong ligament at the wrist into the palm of the hand, where they split up for attachment to the various joints of the fingers. The extensors on the back of the forearm and wrist are arranged in a similar manner.

160. Muscular and tendinous layers attached to the lum-

bar vertebræ, the hip-bones, and the ribs close in the *abdominal cavity* in front and protect its contained organs. The *pelvis* or floor of this cavity is formed mainly by the two hip-bones, the upper margins or crests of which can be felt curving from a sharp point in front, just above the outer part of the flexure of the groin, outward and backward toward the sacrum, which is wedged in between them like the centre stone of an arch [page 102]. The *pelvis* contains the bladder and the lower part of the intestine, over which are packed away the mass of the intestines, the kidneys, liver, spleen, stomach, etc.

161. These organs are separated from the contents of the chest by the *diaphragm*, a thin but strong muscular and tendinous partition attached to the inner surface of the circumference of the lower part of the chest. It forms the floor of the chest or thoracic cavity and the roof of the abdominal cavity. When the fibres contract, the tendency of their action is to stretch the diaphragm tightly like a drumhead between the two cavities, but when they relax, as after an expiration, the partition bulges upward into the chest. On account of this invasion of the thorax by the dome-like convexity of the diaphragm, some of the abdominal organs are situated within the lower part of the cage of the ribs. The *liver* lies usually within the lower ribs and cartilages of the right side. If the middle finger of the left hand, laid flat along one of these ribs, be struck sharply and perpendicularly with the tips of the fingers of the other hand, a dull or flat sound will be elicited as compared with the resonant sound yielded by similar *percussion* over the higher parts of the chest where the air-filled lung instead of the solid liver underlies the finger. When a person lies on his back, the pressure of the other abdominal organs forces the liver wholly within the ribs; in the upright or sitting posture its lower edge may be felt just below the ribs, and if a deep-drawn breath be taken it will be farther depressed. The *stomach* occupies the greater part of the

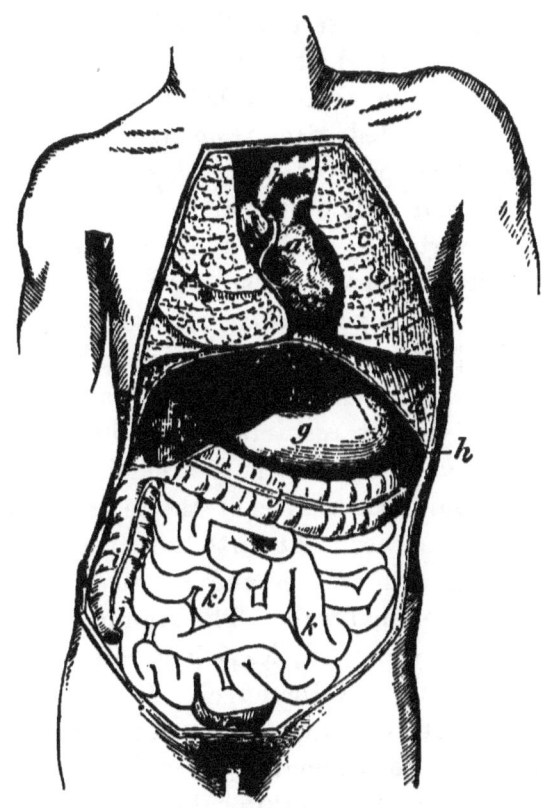

View of thoracic and abdominal organs; anterior walls removed, but the relative position of the ribs, navel, etc., indicated: *a*, heart; *b*, great vessels; *c, c*, lungs; *d, d*, diaphragm; *e*, liver; *f*, gall bladder; *g*, stomach; *h*, spleen: *i*, ascending colon; *j*, transverse colon; *k*, coils of small intestine; *l*, position of ileo-cæcal valve at junction of small and large intestines; *m*, urinary bladder.

epigastrium, the space below the end of the sternum and between the diverging cartilages of the ribs; and on its left side, covered by the lower ribs, is the *spleen*. The *large intestine* traverses the right side from the fold of the groin to the under surface of the liver, where it crosses the abdomen, between the *umbilicus* or *navel* below and the stomach and spleen above, to the left side, which it occupies in its descending course to its termination. The *small intestine* is gathered into coils which fill the space corresponding with the front of the abdomen below the navel; but the *bladder*, when distended, rises from its position in the pelvis into the lower part of this space. The *kidneys* are attached to the rear wall of the abdomen, one on each side of the lumbar vertebræ, so that their position corresponds externally with the loins. These organs are covered with a fine membrane, the *peritoneum*, similar to the synovial membrane [153] lining the joints, which permits them to glide easily on each other during the constant changes of position incident to the respiratory and other movements.

162. The *umbilicus* or *navel* is the remains of an opening through which the blood-vessels of the fetus communicated with those of the mother. At birth these vessels, constituting the umbilical cord, are tied and cut about an inch and a half from the abdominal wall. The stump withers, and in two or three days drops off, leaving a raw surface which, on healing, contracts into the navel. In rare cases, when the umbilical aperture is large, some portion of the abdominal contents may be accidentally forced through it, forming a soft swelling under the skin called a *hernia* or *rupture*. It is treated by means of an abdominal belt, with a pad of suitable size, to support the weak point.

163. *Inguinal hernia* is a protrusion through the passage by which the vessels and nerves of the testicle communicate with the interior of the abdomen. During straining or violent exertion, something is felt to give way, and a soft swelling is found above and to the outside of the pubes

[164]. It may subside when the patient lies down; but it reappears when he resumes the upright position; coughing communicates a notable impulse to it. In aggravated cases the protruded parts may be quite bulky, extending into the scrotum. When they can be returned into the abdomen by making the patient lie on his back, with his knees raised to relax the abdominal muscles, the hernia is said to be *reducible*. Any manipulation to aid its return must be applied with intelligence and gentleness. To press the protrusion in an upward and backward direction with the hand would merely flatten it against the small aperture through which it had escaped. Gentle pressure should be made on the tumor as a whole, but at the same time the effort of the fingers should be to make the part *that came down last go up first, upward, outward,* and *backward,* through the aperture and along the track of the descent. A *truss*, consisting of a steel spring to go around the lower part of the body, and a pad to close and support the weak point, should be applied as soon as the hernia is reduced. The circumference of the body taken in inches, just below the iliac crests, gives the size of truss that will fit a given case. *Suspensory bandages* are used for support in irreducible cases. Hernial protrusions constitute a grave danger to life when they become inflamed. The unyielding aperture constricts or *strangulates* the swollen parts, and the inflammation ends in gangrene unless the constriction is promptly relieved by a surgical operation. Rupture occurs also, but with less frequency, at the middle of the groin where the femoral vessels [182] pass from the abdomen to the thigh. This is known as *femoral hernia*.

164. The *hip-bones* form a strong arch, supporting the body and giving attachment to the powerful muscles which move the lower extremities. Each consists of three bones

soldered together into one irregularly shaped whole; the *ilium* forms the crest and massive side of the bone; the *ischium* constitutes its under portion, or that on which a person rests in sitting; while the *pubes* joins in front with the corresponding bone on the opposite side to form the pubic arch. About the middle of the outer aspect of this

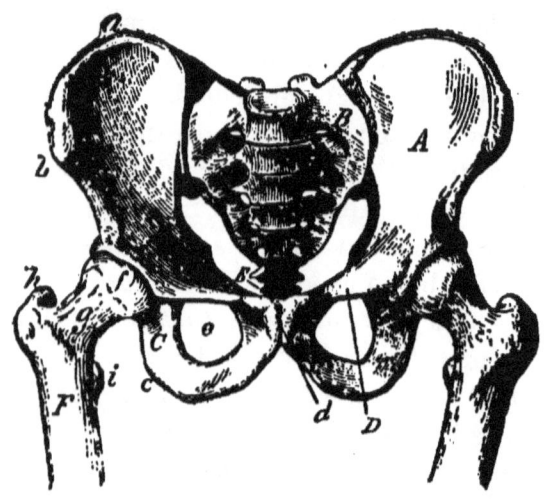

Bony pelvis and upper ends of thigh-bones: *A*, ilium united behind with *B*, the sacrum; *a*, the crest of the ilium, and *b*, its spine; *C*, the ischium, and *c*, its tuberosity; *D*, the pubes, and *d*, the pubic arch formed by the meeting of the two bones; *e*, the obturator foramen or opening, closed in the body by a strong membrane; *E*, three segments of the coccyx, the tip hidden by the pubic arch; *F*, the femur or thigh-bone; *f*, its head in the cotyloid cavity; *g*, its neck; *h*, the great trochanter, and *i*, the small trochanter.

composite bone is a deep cavity, the *cotyloid cavity*, or socket for the head of the thigh-bone. Dislocation is rare at this joint, as compared with its frequency at the shoulder, on account of the depth of the socket, the greater strength of the ligaments forming its capsule, and the power of the overlying muscles.

165. The *femur* or thigh-bone has a strong cylindrical

shaft which expands below into an articular surface and *condyles* [155] at the knee, and above into a round head on its inner aspect, and a large tuberosity on its outer aspect. The head is mounted on a neck by which it is projected upward and inward from the axis of the shaft to the cotyloid cavity. This carries the shaft clear of the hip-bone and gives greater freedom of movement. The tuberosity, the *great trochanter*, can be felt on the outer aspect of the joint, and when the foot or knee is rolled in or out, the trochanter follows the motion.

166. The *knee* is a hinge-joint formed by the lower end of the femur and the upper end of the *tibia* or principal bone of the leg. The massive muscles on the front of the thigh are inserted by a single tendon into the tibia just below the joint; they extend the leg on the thigh. Intimately connected with this tendon is a roundish bone, the *patella* or *knee-cap*, which protects the joint in front. The muscles on the back of the femur are inserted by tendons on each side of the posterior aspect of the head of the tibia, constituting the outer and inner *hamstring* muscles; their contraction flexes the leg.

167. The shaft of the tibia is three-sided, presenting one of its angles along the front of the shin. At its lower end it expands into an articular surface for the ankle-bone, which it encloses and protects on its inner side by a projection called the *internal malleolus*. The *fibula* or outer bone of the leg is much smaller than the tibia, to which it is applied as a support or splint. Its upper end or head is attached to the head of the tibia, where it may be felt just below the outer aspect of the knee-joint. Its lower end forms the *external malleolus*, which encloses and protects the ankle-bone on its outer side. The muscles of the calf are inserted by a strong tendon, the *tendo Achillis*, into the

point of the heel; they extend the foot at the ankle-joint. The muscles beneath those of the calf become tendinous in their lower part and pass under a ligament, between the inner malleolus and the heel, to the sole, where they split up into the *flexor* tendons of the toes. The *extensors* stretch from the outer and fore part of the leg under a retaining ligament in front of the ankle and along the upper surface of the foot to the toes.

168. As the upper extremity is terminated by carpal, metacarpal, and phalangeal bones, arranged as a prehensile organ, the lower is terminated by analogous bones, *tarsal, metatarsal,* and *phalangeal,* modified to secure strength and pliability. There are seven tarsal bones. The *astragalus* or ankle-bone is rounded on its upper surface, which is received into the concavity formed by the end of the tibia and the two malleoli. It is supported on the *calcaneum* or heel-bone; and these two articulate with the others which, with the long metatarsal bones, complete the arch of the foot. The phalanges are in general similar in their arrangement to those of the fingers [158].

CHAPTER II.

THE SYSTEM OF ORGANIC LIFE.

169. THE BLOOD is the essential part of the system of organic life; the accessories are the organs which elaborate the blood, those which distribute, and those which purify it. Every portion of the body, from the bony framework up to the delicate cerebral tissues, which give a material home to the INTELLIGENCE itself, depends on the supply of blood for its growth and well-being.

170. The blood consists of a colorless liquid *serum, plasma,* or *liquor sanguinis,* and a vast number of microscopic cells or *corpuscles* which give the liquid its red color. The serum contains dissolved in it *albumen,* similar to that constituting white of egg, *fibrin,* which, under certain conditions, coagulates spontaneously, and some inorganic salts. When blood is drawn from the body its fibrin consolidates into a soft clot which entangles the corpuscles and serum in its meshes. After some hours the clot becomes smaller and firmer as the coagulated fibrin squeezes out the greater part of the serum which was at first enclosed in it. If the coagulation takes place slowly, affording time for the corpuscles to settle a little before becoming fixed in the consolidating material, the upper surface of the clot may be covered with a *buffy coat,* a grayish-yellow layer of comparatively pure fibrin. Blood sometimes coagulates within the body, as when an artery is tied [388].

171. The cells which give color to the blood are called the *red corpuscles.* They are circular flattened discs, yellowish and translucent, each like a minute, double-concave lens. They consist mainly of a substance called *hæmoglobin*, which combines readily with oxygen, thus enabling them to fulfil their mission, which is to carry oxygen to the tissues for the oxidation and removal of used-up material.

172. White or colorless corpuscles are also found in the blood, one for every four or five hundred of the red corpuscles. They are somewhat larger than the latter and contain granular matter. In shape they are usually spherical, but when closely watched they may be seen to elongate their substance in one or more directions; and as the *amœba*, one

Blood-corpuscles: *a*, red corpuscle, full view of one side; *b*, profile view; *c*, common mode of aggregation during clotting—like piles of coin; *d*, a red corpuscle corrugated, probably old and ready to break down; *e*, amœboid movements of a white corpuscle; *f*, a spherical colorless corpuscle.

of the lowest forms of living matter, moves in this way, these movements are said to be *amœboid*. They have the power of penetrating the walls of the minute blood-vessels; and as they are always present in large numbers when fibrin coagulates in the tissues, they are supposed to be concerned in the coagulation.

173. When the serum of the blood escapes through the walls of the small blood-vessels, as a liquid through filtering-paper, and accumulates in the tissues or cavities of the body, it constitutes what is called an *effusion*. When the white corpuscles penetrate the walls and the effused serum becomes coagulated, the solidified matter is called an *exu-*

dation. The red corpuscles escape only through accidental breaks, and their presence in the tissues is known as an *extravasation* of blood.

174. THE CIRCULATION.—The living body is in a state of constant change. There is a popular belief that the body is renewed every seven years; but the finger-nails require only a few months for their complete renewal. Every organ undergoes change from day to day by the removal of some of its used-up tissue and its replacement by fresh materials. A regiment may have now about the same number of officers and men that it had a few years ago, but the individuals that compose its numerical strength are not the same; those lost by discharge and death have been replaced by recruitments. So the organs of the human body are subject to processes of disintegration on the one hand, and renovation by the blood on the other. In every movement some minute particles of the muscles become unfit for further service and have to be replaced; in every thought or mental impulse some part of the nervous system becomes used up, as the plates of a galvanic battery are dissolved to develop its power. To effect the repair of this constant waste a constant circulation of the reparative material is needful; but although this is provided by nature, the activity of the destroying influences is so great that recurring periods of rest or sleep are required to permit of the perfect recuperation of the tissues.

175. The circulation of the blood is effected by the *heart*, which drives it through the system; the *arteries*, which conduct it to the tissues; the *capillaries*, from which the reparative operations are conducted, and the *veins*, which gather up the blood and return it in an altered and impure condition to the heart. Before the blood starts again on its *systemic* round it is driven to the lungs, where it becomes purified. This accessory to the systemic circulation is called the *pulmonary circulation*.

176. To keep up these two currents of blood, the heart is divided by a partition into two sides or chambers, the left receiving and delivering *pure* or *arterial* blood, the right receiving and delivering *impure* or *venous* blood. These chambers are formed of muscular walls, which by their contraction drive the blood into the vessels, as the

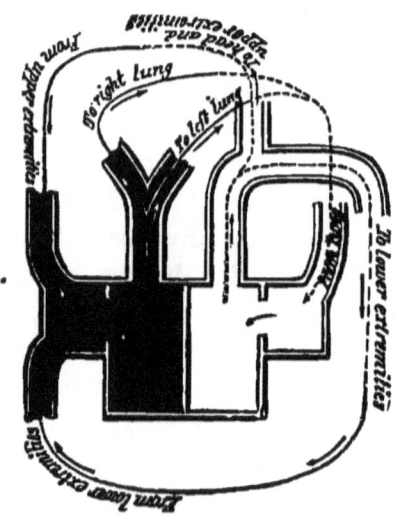

Diagram of the circulation.

hand, by its contraction on the rubber bulb of an atomizer, drives the air through its tube. The upper part of each side forms a receiving-chamber or *auricle* for the venous blood on the one side and the arterial blood on the other. They become filled while the lower part of each, the *ventricle* or delivering-chamber, is contracted in the act of sending out its charge of blood. As soon as each ventricle is emptied it relaxes, receives into its cavity the accumulated blood from the auricle, and again contracts to drive out this fresh supply [262]. The two sides of the heart act simultaneously, so that at the ventricular contraction the arterial blood on the left rushes off to the tissues, and the venous on the right to the lungs. To prevent a reflux into the auricles at this time, the communication between the chambers is guarded by *valves ;* and to prevent a return from the vessels during the subsequent relaxation of the ventricles, the mouths of the former are similarly guarded. The left ventricle is thicker and stronger in its muscular walls than the other chambers, as it has to drive the blood to the uttermost parts of the system. The discharging capacity of each ventricle is about two fluid ounces.

177. The heart is conical in shape, and about the size of the closed fist of the individual. It is imbedded between the lungs, and sustained in its position by the many large vessels which connect with its upper part. A strong fibrous membrane, the *pericardium*, forms a loose sac around it; and as the interior of this sac and the exterior of the heart are both coated with a smooth serous membrane, the movements are effected without friction. The heart lies behind the middle of the sternum, extending from the central line of the body about three inches toward the left, but only half that distance toward the right. Its lower end or apex corresponds with a point two inches below the left nipple and one inch to its inner side (page 99). When the ear is laid over the *cardiac region* the heart-sounds are heard. They have been likened to the pronunciation of the syllables *lŭbb-dŭp*, and are separated from the next following repetition by a well-marked pause. The first sound corresponds with the contraction of the ventricles and the rush of blood through the great arteries; the second is chiefly caused by the sudden closing of the valves in these arteries

when ventricular relaxation begins; and the pause corresponds with the passage of the blood from the auricles into the relaxing ventricles.

178. The great artery of the systemic circulation, the *aorta*, begins at the upper part of the left ventricle, and, passing upward behind the sternum for a short distance, curves backward to the left side of the vertebral column. From the upper aspect of the arch arise the vessels which supply the head, neck, and upper extremities.

179. The *carotid* arteries run upward in the neck on each side of the trachea and larynx, underneath the inner margin of the sternomastoid muscle, giving off large branches for the neck and face, and a large vessel, the *internal carotid*, for the interior of the cranium. One of its branches, the *temporal*, may be felt pulsating in front of the ear, just above the articulation of the lower jaw, and a branch of this, the *anterior temporal*, may often be *seen* pulsating as it crosses the temple on a line from the upper border of the ear to above the orbit.

180. The *subclavian* arteries run upward and outward from the arch of the aorta behind the clavicle and over the upper and outer surfaces of the first rib to the *axilla* or armpit. The *axillary* artery is the continuation of the subclavian as it passes along the inner side of the shoulder-joint and upper part of the shaft of the humerus. The artery below the insertion of the pectoral muscles [153] is called the *brachial;* it lies along the inner margin of the biceps muscle, passing from the inner side of the humerus in its upper part to the front of the bone at the elbow-joint, where it divides into two branches, the *radial* and *ulnar*. These branches descend the forearm, overlapped by muscles, but near the wrist they become superficial and are easily detected by the fingers. The radial is the vessel usually selected for ascertaining the rate and other characters of the arterial *pulse*. On reaching the palm the two vessels reunite over the metacarpal bones, and from the arch formed by their junction branches descend between the bones, breaking up at the clefts into terminal branches which course along adjacent sides of the fingers to their tips.

181. After completing its arch the aorta descends in the thorax along the left side of the spinal column until, opposite the fourth lumbar vertebra, it breaks up into two large vessels, the *common iliac arteries*. Each of these extends downward and outward into the pelvis, where a large branch, the *internal iliac*, is given off to the contents of the cavity, its walls and the muscles connected with the outer surface of the ilium. The continuation of the main trunk, under the name of *external iliac*, reaches the middle of the groin, where it enters the thigh and becomes known as the *femoral artery*.

182. The pulsation of the femoral may be discovered by the fingers from the middle of the groin downward for a short distance. It then dips under the muscles and becomes lodged in a canal of fibrous tissue along the inner side of the bone. At the junction of the middle with the lower third of the thigh it passes backward into the *poples* or ham, where under the name of *popliteal* artery it descends to below the knee-joint, where it divides into two tibial branches. The *anterior tibial artery* penetrates to the front of the leg between the heads of the tibia and fibula, descends, covered by muscles, along the outer side of the tibia, and under the name of the *dorsal artery* of the foot forms an arch from which branches are distributed to the phalanges. The *posterior tibial* descends along the back part of the leg, becoming superficial below, where it passes between the internal malleolus and the tendo Achillis [167] to

reach the sole. Here it forms the *plantar arch* for the supply of the toes.

183. Each of the various arteries that have been mentioned gives off branches to the parts in its neighborhood. These divide and sub-divide until they reach a size but little larger than the capillaries in which they terminate. The small arterial vessels form a network or vascular framework, in which the tissues are bedded like the cellular substance of a leaf between the meshes of the veinlets derived from the leaf-stalk.

184. The walls of an artery are so strong and elastic that when empty they do not collapse, but remain open like a rubber tube. They consist of an outer coat of tough elastic tissue, a middle layer of muscular and elastic tissue, and a smooth interior lining. When the heart throws its charge of blood into the arterial tubes the walls of the latter yield to the dilating force, but this yielding is momentary; the contractility of the muscular coat of the vessels, closing on the contained blood, forces it onward to the tissues, back flow being prevented by the valves in the aorta.

185. An organ or tissue does not require the same quantity of blood at all times. The brain when actively engaged requires more than when asleep; a muscle in active use requires more than when it is at rest. An increase in the action of the heart provides more liberal supplies when the requisition is made by the system as a whole; but local requirements are supplied by the contractility of the arteries under the superintendence of the nervous system. When an organ requires an increased supply its arteries become larger by the relaxation of their muscular fibres. Congestions or local determinations of blood [305] in cases of injury or disease are effected by this action of the muscular coat. Nervous impressions often operate on the size of the vessels, as when the face becomes pale or flushes under the influence of certain emotions. The contractility of the middle coat has also an important bearing on the suppression of hemorrhage [386].

186. The expansion of the arterial tubes, consequent on the sudden delivery into them of the contents of the left ventricle, constitutes the *pulse-wave* as felt at the wrist or elsewhere. The average pulse in the adult is about 70 or 75 per minute; in children it is more rapid; in infants, 110 or 120. A *natural* pulse is of normal frequency, *equal* or *regular*, that is, having all its pulsations similar and at equal intervals, and neither *hard* like a cord nor *soft* or easily obliterated, yet susceptible of a certain degree of impressibility by the finger. As regards the number of beats per minute, the pulse is said to be *slow, frequent,* or *rapid;* the term *quick pulse* has no reference to frequency, but to suddenness of impulse. A pulse is *irregular* when its beats lack uniformity in strength or intervals; *intermittent* when one beat is omitted after a certain number of regular beats. A *full* pulse is one of large calibre; a *small* pulse is generally rapid and *weak* or *feeble,—thready* or *wiry* if it be hard, and *fluctuating* if soft. A *febrile* pulse is rapid, full, and hard if the patient be robust; rapid and small if he be prostrated. Exercise increases the frequency of the pulse. Its beats are more frequent when one is sitting or standing than when lying. Weak pulses suffer greater acceleration by exercise than strong ones.

187. If a part be deprived of its supply of blood for a certain length of time, it will mortify. To lessen the danger from accidental obstructions, the branches of one artery communicate or *anastomose* with those of another. When an artery, such as the axillary, has been wounded and tied, the blood supply for the limb is kept up by

anastomosis. The direct route being blocked up, the current passes through one or more of the branches given off above the ligature, and these, by their connections with those given off below the ligature, establish a *collateral circulation* by which the parts below the obstruction are supplied with blood. Anastomosis is more extensive in the upper than in the lower extremity, and hence gangrene after injury is less common in the hand than in the foot. Where the anastomosing branches are large and the collatteral current readily established, there is danger of the recurrence of bleeding from the lower or far end of a divided artery, unless that end as well as the upper or near one be closed by a ligature.

188. As the arteries subdivide into smaller branches their walls become thinner, until they end in a vast number of minute tubes which surround the elementary cells or fibres of the various tissues with a freely anastomosing vascular network. The vessels of this, the *capillary* system, are just large enough to permit the passage of the red corpuscles in single file. Their delicate walls permit the plasma to exude into the tissues for the processes of growth and repair; the red corpuscles yield up the oxygen which they have brought from the lungs for the oxidation or combustion of used-up materials; heat is developed, and the liquor sanguinis dissolves and washes back into the vascular current the carbonic acid and other more complex matters which result from the oxidation. These changes are manifested in the blood by that darkening of its color which distinguishes venous from arterial blood.

189. The *veins* collect the blood from the capillaries and carry it back to the heart. The venous system is more capacious than the arterial, for many veins have no corresponding arteries, and all the smaller and some of the large arterial vessels are accompanied by two returning veins. The radial, ulnar, and brachial arteries, and those of the leg, have each two companion veins; but from the axillary and popliteal onward to the heart the arteries have but one associated vein.

190. The veins of the upper extremity unite into a single trunk, the *axillary* vein, which, on crossing over the first rib behind the clavicle, becomes the *subclavian* vein. Behind the sternoclavicular articulation the subclavian unites with the *internal jugular*, the companion of the common carotid artery, which returns the blood from the head, face, and neck. The trunks formed by these two large veins join behind the sternum, forming the *superior vena cava*, which terminates in the upper part of the right auricle (page 119).

191. The veins of the inferior extremity unite into a single trunk, the *popliteal*, which becomes successively the *femoral*, the *external iliac*, and the *common iliac* vein. By the junction of the common iliacs is formed the *inferior vena cava*, which runs upward on the right side of the aorta and ends in the right auricle.

192. The inferior cava, in its passage upward along the spine, receives blood from the kidneys, the testicles, and the abdominal walls; but the blood from the organs of digestion passes through the liver before entering the direct channel to the heart. The veins from the intestines, the stomach, and the spleen unite into one large vessel, the *portal* vein, which enters the liver, subdividing, like an artery, into smaller branches, and ultimately into a capillary system, the blood of which is gathered up by a vein, the *hepatic*, and passed into the inferior cava as the latter vessel passes upward behind the posterior border of the organ. There are thus in the liver two series of blood-vessels—in fact, two circulations; the ordinary arterial circulation, ending in the hepatic vein, and intended for the nutrition of the organ; and the portal circulation, also ending in the hepatic vein, but intended apparently for the exposure of the venous blood of the digestive organs to some special action prior to its admission into the general circulation. As the veins of the portal system contain much crude material absorbed from the stomach and intestines, it may be inferred that the object of passing their blood through the liver is to prepare the crude material for future use as a part of the mass of the circulating blood.

193. The veins which have no corresponding artery are usually superficial in position; and as the object of their existence appears to be to furnish a route for the returning blood when the deeper veins are temporarily occluded [195], their intercommunicating branches are numerous and large. The *external jugular* returns the blood from the exterior of the head and certain parts of the face; its branches unite to a single trunk, which extends from just below the ear, downward, across the sternomastoid, and then parallel to the posterior border of that muscle, to the middle of the clavicle, where it ends in the subclavian vein.

194. The superficial veins on the back of the fingers and hand, and on the ball of the thumb and little finger, become aggregated into three sets of vessels. Those on the outer side of the forearm unite near the elbow to form a large vein, the *cephalic*, which courses up along the outer border of the biceps muscle to join the axillary vein. Those on the inner side form the *basilic* vein, which continues along the inner border of the biceps to join the brachial veins. Those on the front of the forearm unite to a single trunk, the *median*, which runs upward to the bend of the elbow, where it divides into two short branches, the *median cephalic*, extending obliquely outward to join the cephalic, and the *median basilic,* extending inwards to the basilic. In former times when bleeding was extensively practised, the outer of these communicating veins was usually selected for the operation, as the inner one lies immediately over the brachial artery, just above its point of division into the radial and ulnar. The *external saphenous* vein collects blood from the upper and outer part of the foot and the posterior aspect of the leg, and discharges into the popliteal. The *internal saphenous* vein passes along the inner aspect of the

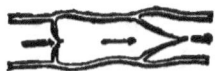

Valves of veins: The arrow indicates the direction of the flow toward the heart.

leg and the middle of the thigh to a little below the groin, where it joins the femoral vein.

195. The walls of the veins are so thin that when filled the color of their contents shows through them, and when empty they collapse. Slight pressure closes them and obstructs their circulation. They have none of the contractility or elasticity that characterizes the arterial walls. After death the arteries are emptied by means of this contractility, but the veins are found to contain blood. The lining membrane of the veins is pinched up at various points into folds which act as *valves,* permitting the passage of blood toward the heart, but overlapping each other and forming a partition across the tube when any pressure tends to force the current back on the capillary system. When a muscle contracts, the veins in its neighborhood are subjected to a pressure which obliterates their channels; but owing to the free communication of their branches, the blood escapes into col-

lateral veins; and this escape, by virtue of the system of valves, is always in the direction of the heart. Muscular movements thus accelerate the venous current. The heart operates automatically. When the ventricles are filled they contract to discharge the blood, and then relax to be filled again. In the healthy condition the rapidity of the heart's action depends on the inflow from the veins. Exercise calls for more blood to the muscles, but this same exercise hastens the return of the venous current and enables the heart to meet the demand. When, for instance, one is rowing, the deep veins are alternately filled and emptied by muscular action at each sweep of the oars—filled from the capillaries, and emptied in the direction of the heart through the superficial veins.

196. As the walls of the superficial vessels have little support from adjoining textures, they sometimes yield to the pressure of the contained blood, and become permanently dilated or *varicose*. The internal saphenous vein and its branches are often thus affected, becoming large, tortuous, and knotty. Some long-continued obstruction is generally concerned in the causation, as the pressure of tight garters or that of constipated bowels on the venous trunk within the pelvis. The benefit to be derived from elastic stockings in such cases is obvious. Bleeding from the rupture of varicose veins should be restrained by compresses with a firmly applied bandage. *Hemorrhoids* or *piles* are enlargements of the veins around the anus and lower part of the bowel, due to the pressure of constipated bowels or to conditions of the liver, which obstruct the passage of the blood through the portal system [192].

197. The venous blood returned to the right auricle by the superior and inferior venæ cavæ, is retained in that chamber only until the relaxing ventricle opens to receive

it. The distention of the ventricle is immediately followed by its contraction, which drives the blood into the *pulmonary artery*, the branches of which divide and subdivide until they reach the terminal subdivisions of the air pas-

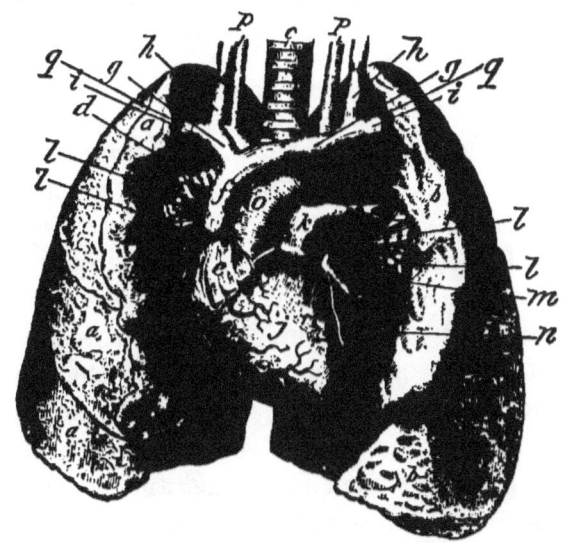

THE THORACIC ORGANS: *a*, right lung divided into three lobes, and *b*, left lung into two lobes, their anterior margins thrown back to expose the deeper parts; *c*, trachea with its cartilaginous rings; *d*, right bronchial tube; *e*, right auricle, receiving from above *f*, the superior vena cava, which is formed by the junction of *g*, *g*, the right and left innominate veins, and each of these by the confluence of *h*, *h*, *i*, *i*, the jugular and subclavian of its own side; *j*, the right ventricle, giving issue to *k*, the pulmonary artery, which divides into two branches, that for the right lung passing behind the other great vessels; *l*, *l*, *l*, *l*, pulmonary veins, bringing oxygenated blood to *m*, the left auricle; *n*, the left ventricle, from which the blood is carried to the organs and tissues by *o*, the aorta; *p*, *p*, *q*, *q*, carotid and subclavian arteries, given off from the arch of the aorta.

sages, where they end in a capillary system which surrounds each air cell with a vascular network. Here the blood undergoes changes which are the reverse of those which take place in the systemic capillaries. The red corpuscles exhale carbonic acid and absorb oxygen, and, as a result, the blood loses its dark color and becomes a florid red; heat is carried off by the watery vapor and the accompanying exhalations [243]. The oxygenated or arterialized blood is gathered up by veins which complete the circuit by joining the left auricle.

ALIMENTATION.

198. As the blood is subjected to a continuous loss of its nutritive materials in the capillaries of the systemic circulation, provision is made by nature to supply its elementary principles. The materials of the blood are obtained from the substances ingested as food. These are divided into three classes—the nitrogenous, the non-nitrogenous, and the inorganic. The *nitrogenous* contribute the albuminous and fibrinous principles for the repair of the muscular system. They include the albumin of fresh meat, the casein of milk, the gluten of flour [574] and analogous principles in other vegetables as peas, beans, mushrooms, etc. The *non-nitrogenous* replace those elements of the tissues that are destroyed by oxidation during the production of force and the development of animal heat. They include the fats, starches, and sugars, all of which consist of carbon and hydrogen in a readily oxidizable form. Taken in excess of the immediate wants of the system, they are stored up as surplus fat, to meet the emergencies of sickness or impoverished diet. The *inorganic* consist of various salts which are usually found already prepared by nature in the substances that are used as food. In wheaten flour there is nearly one per cent. of mineral matters, which are just those that are required by the human system; in fresh meat and all other alimentary substances there is a similar supply of inorganic matter—*common salt* is perhaps the only one of this class which is added specially to the diet. To these three classes a fourth is sometimes added, comprising such articles as have been called *accessory foods*—tea, coffee, cocoa, alcohol, pepper, and other condiments; but these are stimulants rather than foods. Water is not usually regarded as a food, although it enters the system along with the food and forms so large a percentage of the various tissues of the body. Death occurs from its deprivation more quickly than from the deprivation of food; men have fasted for many weeks, but they will die for want of water in as many days.

199. Each of the natural substances used as food contains a proportion of all the elementary principles that are needful to the formation of the blood supply. Wheaten flour contains 10 to 15 per cent. of gluten, 60 of starch, 7 of sugar, and over 1 each of fatty and inorganic matters; fresh meat yields 20 per cent. of albuminoids, varying proportions of fat, and nearly 2 per cent. of salts; pork, 10 per cent. of albuminoids and a larger proportion of fat.

200. If any one of these substances contained the elements in the proportion in which they are required by the human body, that substance would be a perfect food. Milk is a perfect food for the child. It contains nitrogenous matter in the form of casein, fat globules in the cream, and *lactose* or sugar of milk and salts, chiefly phosphate of lime, dissolved in its water. The requirements of the infant do not vary from day to day, and the one food suffices for its growth and development; but as it grows older, varying conditions of exercise or repose call for more or less of the nitrogenous elements, and varying conditions of external temperature for more or less of the non-nitrogenous. Milk then ceases to afford the proportions fitted for all the conditions in which the individual may be placed, and articles

of diet must be selected which in their totality give the proportions required by the system. Great exertion calls for a larger proportion of nitrogenous, great cold for a larger proportion of fatty matters, while the starches and sugars suffice for the wants of the system in the absence of these special calls. Variety in diet, which is generally regarded as a matter of taste, originated in the necessities of the system.

201. The total quantity of food necessary for support is also measured by the conditions affecting the body. A man who passes the day in hard labor or active exercise requires that *full diet* of which the army ration is an illustration. If he be in hospital with some chronic discharge, which acts as a drain upon the system, a full diet is also needful. Ordinarily a patient in hospital requires much less, particularly of the fatty elements, than is allowed by the army ration; and the money value of the unused portions of the ration constitutes the hospital fund.

202. When an *excess of food* is taken, it is not absorbed, but accumulates in the intestinal canal, undergoing changes which tend to putrefaction. The acrid matters thus produced are taken up into the blood and induce a febrile condition, or they irritate the intestinal canal and cause a diarrhœa which brings about their expulsion.

203. A *deficiency of food* occasions a feeling of faintness or sinking at the epigastrium, gnawing pain in the stomach, headache, and weakness. Want of food is a common and often unsuspected cause of sleeplessness. Patients in hospital should never be dosed with opium when a cup of beef tea with a cracker will remedy their complaint. When the deficiency of food is continued, progressive weakness and emaciation follow with increased liability to disease and a special tendency to the scorbutic condition [77].

204. *Mastication and Swallowing.*—When the pharmacist desires to extract the active principle from some root or tough vegetable product, he slices, beats, or breaks it into small portions that his menstruum may penetrate it and dissolve out the valuable material. So nature operates in dealing with the crude materials from which the elements of the blood are to be extracted. To facilitate their reduction a liquid, the saliva, oozes from its secreting glands into the mouth, to be incorporated with the triturated food which it transforms into a soft pultaceous mass. The mouth, *fauces* or throat, and *œsophagus* or gullet, are lined, like the rest of the alimentary canal, with a membrane, which secretes a thick viscid liquid called *mucus*, which lubricates their interior, and enables the mouthful to glide easily into the stomach.

205. If mastication be not thoroughly performed, the stomach suffers in the long run. Those who would be free from dyspepsia should have good teeth and make good use of them. Nature provides man with two sets of teeth. In the *milk* or *temporary set* there are ten teeth in each jaw; four *incisors*, or sharp-edged cutting teeth, in front, with one sharp-pointed *canine*, or tearing tooth, on each side of them, and two broad-crowned *molars*, or grinding teeth, on each flank of the dental line. The central incisors of the lower jaw appear about the sixth or seventh month, those of the upper set closely following; the lateral incisors of the lower about the tenth month, followed by the corresponding upper teeth; the anterior molars shortly after the end of the first year, the canine or *eye-teeth* at a year and a half, and the posterior molars during the third year. Teething is sometimes attended with considerable local irritation, restlessness, loss of sleep and appetite, and even symptomatic febrile action; diarrheal troubles occasionally alternate with inflammatory conditions of the skin covering the buttocks or scalp, and sometimes convulsions are developed. Relief may often be afforded in such cases by lancing the gums when it is evident that the tooth is close to the surface.

206. The permanent teeth are sixteen in number in each jaw: *Four* incisor and *two* canine teeth supplant those of the temporary set; *four small* or *false molars,* two on each

Central incisor of upper jaw.

Canine or eye-tooth of upper jaw.

Second bicuspid of lower jaw.

side, called also *bicuspids,* because they have two points or cusps on their crown, replace the molars of the child; and *six large* or *true molars,* three on each side, called also *multicuspids,* because they have several points on their crown to facilitate the grinding of food, make their appearance behind the bicuspids. The lower teeth erupt before the corresponding teeth of the upper jaw; the first or anterior molars about the sixth year; the central incisors at the seventh; the lateral incisors at the eighth; the first bicuspids at the ninth, and the second at the tenth year; the canines from the eleventh to the twelfth year; the second or middle molars at the thirteenth, and the third or posterior molars, the *wisdom teeth,* as they are called, from the seventeenth to the twenty-first year, or later. Sometimes young soldiers suffer considerably from pain and swelling about the back part of the alveolar arch, due to the incoming of a wisdom tooth. When the swollen gum overlying the crown of the tooth is cut across, the symptoms will speedily subside.

207. Each tooth is divided anatomically into the *crown*, which is visible in the mouth; the *neck*, which is embraced by the gum; and the *root*, hidden by the gum and socketed in the jawbone. The root of each of the incisors, canines,

  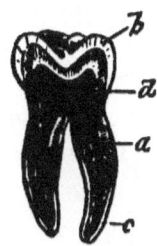

Second molar of upper jaw.   Wisdom tooth of upper jaw.   Section of lower molar: *a*, dentine; *b*, enamel; *c*, crusta petrosa; *d*, pulp cavity.

and bicuspids is single and conical, the canine fang being longer than the others. Each lower molar has two roots, one in front and one behind, and each upper molar three roots, two toward the outer and one toward the inner aspect of the jaw; but the fangs of each wisdom tooth are usually consolidated into one which presents grooves or markings of a division into three in the upper and two in the lower jaw.

208. The teeth are composed of a substance called *dentine*, somewhat harder than bone. This is crowned with a dense white enamel and coated on the fangs with a thin layer of true bone, which is here called *crusta petrosa*. In the interior of each tooth is a cavity containing the *dental pulp*, which consists of nerves and blood-vessels that enter by an aperture at the end of the fang.

209. Decay takes place by the softening and breaking

down of the dentine, at first into superficial cavities, which afterward lay open the dental pulp and, undermining the enamel, permit the caving in of the crown and the total destruction of the tooth. The teeth may be preserved for many years by the use of the tooth brush and by having all cavities filled with gold or other durable filling at the hands of a dentist. Speedy decay follows neglect in all cases, although some teeth and some individuals are more liable to suffer than others. The anterior molars give way soonest, and are frequently decayed before the posterior molars have erupted. The pain associated with this decay is sometimes so severe in its paroxysmal exacerbations as to call for emergency treatment [346]. A crust of *tartar* accumulates on those parts of the teeth that are not subjected regularly to the use of the tooth brush. This deposit, which consists of a cement of lime and organic matter, adheres so firmly that it has occasionally to be scaled off by a sharp steel instrument [350]. The margin of the gum behind the front teeth of the lower jaw and the outer surface of the posterior molars are favorite sites for its accumulation.

210. The *saliva* contains a fermentative principle, *ptyalin*, which, when mixed with starch, changes that substance first into dextrin and then into glucose [564]. The change is immediate, so far as regards a small percentage of the starch, but the fermentation is brought to an end as soon as the mouthful reaches the acid liquids of the stomach.

211. The *stomach*, when distended, is a somewhat pear-shaped bag occupying the epigastrium, its large end on the left under the diaphragm, its smaller or *pyloric* extremity on the right under the liver. [See illustration on page 99.] Its walls consist of flattened bundles of longitudinal, oblique, and circular muscular fibres, with an external peritoneal [161] covering and an internal mucous lining. Embedded in the last are many minute tubular follicles from which, when required for digestion, the *gastric juice*, an acid liquid containing a ferment, *pepsin*, oozes into the stomach, as the perspiration sometimes oozes from the pores of the skin or the saliva into the mouth during mastication [262]. When a meal is taken, the gastric juice is freely secreted and incorporated with the masticated food by the contraction of the muscular walls of the stomach. Starch and fat are unchanged in the stomach except for mechanical division, but albuminous materials become dissolved into a sour-smelling liquid called *pepton*, which is in part absorbed by the veins of the stomach and in part escapes at the pylorus along with the *chyme* or grumous liquid which the stomach transfers to the intestine as the result of its digestion.

212. The process of liquefying the albuminoids, which is the object of gastric digestion, occupies from one to four or five hours, according to the quantity and digestibility of the food and the efficiency of the preliminary mastication. It can easily be understood why a mass of tough gluten and unbroken starch cells, such as exists in sodden, unraised bread will cause a sense of heaviness or oppression in the stomach for some time after it has been eaten. Meats that have been preserved in brine become hard and difficult of digestion. An excess of fat also interferes with the action of the gastric juice and prolongs the stay of the food in the stomach. Veal and lamb require more time than beef or mutton. Peas and beans, on account of their large proportion of nitrogenous elements, make a greater call on the powers of the stomach than the cereal grains. Poultry, fish, and oysters are quickly digested.

213. The contractile walls of the stomach keep it in close

128   THE INTESTINES.

contact with the food during the progress of digestion. The pyloric end, leading into the intestine, is guarded by a circular band of muscular fibres which permits liquids to drain away, but prevents the passage of undigested food. In the healthy condition there is no tendency to regurgitation through the œsophageal opening, but in certain morbid states of the stomach its contents are thrown out by *vomiting*.

214. The *small intestine* is a long narrow tube in which the chyme is treated after its passage from the stomach. It measures about twenty-five feet from its commencement at the pylorus, under the cartilages of the ribs of the right side, to its termination in the large intestine in the right iliac region just above the fold of the groin. [See page 99.] It is gathered up into coils which are held in position by folds of a membrane called the *mesentery*. The walls consist of a muscular layer, covered with peritoneum externally, and lined with a thick mucous membrane, arranged loosely in many pleats or folds to expose a greater surface to the material which passes over it. The mucous membrane, besides secreting mucus for the protection of itself against matters of an acrid nature, is provided with many small follicles which aid the liver and pancreas in the digestion of starchy substances [218].

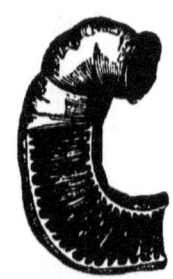

Folds of the mucous membrane, *valvulæ conniventes*, of the small intestine.

215. The spleen lies under the lower ribs of the left side, close to the stomach. It is about the size of the fist; but as it is elastic and distensible its size depends chiefly on the quantity of blood which it contains. One of its functions is that of acting as a safety valve to the blood current for the supply of the digestive organs. When full and actively engaged these organs require a larger blood supply than when empty; and as the spleen is always smaller when the stomach is full than when it is empty, it is supposed to be a distensible cistern which serves to accommodate the blood when that liquid is not required for active operations in the stomach and intestines.

216. The *liver* [page 99] secretes a dark-greenish liquid, the *bile*, which is stored, when not required for use, in a small pear-shaped receptacle, the *gall-bladder*, opening into the intestine near the pylorus. This secretion acts as a stimulant to the muscular coat of the intestines, inducing those contractions which carry the contents onward from the upper to the lower end, a movement known as the *peristaltic* or *vermicular* motion of the intestine [262]. The bile is strongly alkaline, saponifying fatty substances, as when oil is shaken up with liquid ammonia or caustic soda, and rendering them soluble and susceptible of absorption. It is also an antiseptic, preserving the chyme from putrefaction during its sometimes slow progress through the intestinal canal.

217. The *pancreas* or sweetbread secretes an alkaline liquid containing a ferment, *trypsin*, which has the same influence on starch that is possessed by ptyalin [210]. The pancreatic juice operates also on the liquid albuminoids that have escaped the action of the stomach, transforming them into pepton; and by its alkalinity aids the bile in reducing the fats to a soluble condition.

218. The chyme diminishes in quantity as it approaches the lower end of the small intestine. The pepton, dextrin, glucose, and emulsified fats are absorbed by the veins of the intestine, but chiefly by a set of vessels called *lacteals*. The mucous membrane of the small intestine is so beset with minute processes or *villi* as to resemble velvet. Each of these processes contains a special absorbing vessel in its centre. The vessels from adjoining villi unite, forming larger vessels, which may be detected as whitish lines on the intestine if digestion is in progress at the time of the examination. The white color is due to minute globules of emulsified fat in the *chyle* or liquid which the vessels absorb. These vessels enter the folds of the mesentery, where they pass through the substance of a number of small bodies, each about the size of an almond. In these, the *mesenteric glands*, a change takes place in the chyle. It becomes capable of coagulating and many white corpuscles like those of the blood are found in it. After leaving the glands, the vessels unite into two or three trunks, which terminate in a large vessel, the *thoracic duct*, by which the chyle is led upward along the front and left side of the vertebral column to the left subclavian vein near the junction of the internal jugular.

219. Digestion and absorption take place in the *large intestine*, for nutrient enemata are changed and taken up by the vessels even from its lower part—the *rectum*. During the passage of the intestinal contents along the upper part of the large intestine, the secretions that were concerned in their digestion in the small intestine are absorbed, leaving only a semi-solid mass of refuse and undigestible material called *fæces*, which is excreted. A valve, the *ileo-cæcal* [page 99], guards the entrance to the large intestine and prevents backflow into the small intestine by any movement of compression. Attached to the head of the large intestine near this valve is the *appendix*, a small rudimentary organ which is sometimes the seat of dangerous inflammations.

220. The blood is thus seen to be supplied with its nutritive material from the albuminoids, starches, fats, sugar, and salts of the food, the crude results of the digestive process finding their way into the circulation by the hepatic vein after filtration through the liver, or by the lacteal vessels after filtration through the mesenteric glands. The lacteals are, however, only a part of a larger absorbent system which pervades the whole body. This, the *lymphatic system*, consists of minute vessels which ramify in the tissues, often following the track of the veins in their junction one with another to form larger vessels. They absorb from the tissues those matters that have been exuded or formed during the nutritive or reparative processes, and return them to the blood for further use as nutritive material, or to be carried to the proper organ for elimination as refuse. Owing to the thinness of their walls and the transparency of the *lymph*, as the liquid which they contain is called, they are made out with difficulty unless injected for the special purpose of displaying them. Sometimes, when inflamed, their tracks may be observed in the living body, as when from a poisoned or festering sore on the hand their red lines are found to extend along the forearm and arm to the armpit. The flow of the lymph is occasioned by muscular movements and valves which, as in the case of the veins, direct the current onward to the heart. The vessels anastomose freely with each other, and at certain parts of their course are connected with small *glandular bodies* which are usually aggregated in particular localities, as the groin, armpit, neck, etc. The glands effect a change in the character of the lymph, increasing its coagulability, providing it with *white corpuscles*, and preparing it for readmission into the blood by way of the thoracic duct. Matters which would poison the blood, if admitted into its current, are retained in the glands, in which they often set up a suppurative inflammation, which leads to their own extrusion from the system.

EXCRETION.

221. Respiration.—The lungs fill the cavity of the chest with the exception of that part in the centre and toward the left occupied by the heart. [See illustrations on pages 99 and 119.] Each is enveloped in a serous membrane, the *pleura*, which lines also the interior of the walls of the chest and the upper surface of the diaphragm, to facilitate the motion of one part on the other. The space between the pleura covering the lung and that lining the walls is called the *pleural sac* or cavity; but in health there is no pleural cavity, for the lung lies in close contact with the walls which protect it.

222. Each lung is composed of the branchings of an air tube and the associated divisions of the pulmonary arteries and veins which carry the blood to and from the organ. The air tube or *trachea* divides behind the sternum into two branches or *bronchial tubes*, one for each lung. To prevent its closure during movements of the neck, the walls of the trachea are formed of narrow cartilaginous rings closely set in a strong fibrous membrane. The ultimate branches of the bronchial tubes end in minute air cells, like grapes on the terminal twig of a vine, each cell surrounded with a network of capillary vessels in which the pulmonary arteries end and the pulmonary veins find their origin. The air passages are lined with mucous membrane covered with cells surmounted on their free ends with microscopic filaments, or ciliæ, which, by a constant swaying motion prevent the secreted mucus from backing into the air cells.

223. An upward and outward movement of the ribs, with

a concurrent downward movement of the diaphragm [161], creates a partial vacuum in the lungs, which is immediately filled by inflow of air through the larynx and trachea, constituting the act of *inspiration*. No sooner is this completed than the diaphragm relaxes, the ribs fall, and the capacity of the chest being thereby lessened, some of the contained air is thrown out, constituting the act of *expiration*. The alternation of these two movements constitutes the act of *respiration*, which, in the normal condition of the system, is performed from fifteen to eighteen times per minute, or at the rate of one respiration for every five beats of the heart. The motive power in quiet respiration is mainly furnished by the action of the diaphragm; when the breathing is more active the muscles which elevate and depress the lower ribs become engaged; in *forced respiration* every muscle bearing on the thorax is called into action, and even those of the face participate in the movement, dilating the nostrils visibly to facilitate the ingress of the air.

224. The organ of voice or *larynx* is situated in the upper part of the air tube. Its walls are here strengthened by large cartilages which increase its size on its outer aspect; but internally, opposite the prominence of the cartilages, the tube is narrowed to a triangular chink, the *rima glottidis*, by two folds of mucous membrane, the *vocal cords*, which vibrate in the passage of the air current between them and produce the voice.

225. The lungs have a capacity of over three hundred cubic inches, but ordinarily the air movement into and out of the chest does not exceed thirty cubic inches. This quantity at each inspiration enters the lungs, mixing with that already there, and at expiration the same quantity of the mixed air is thrown out, the removal and replacement

sufficing to keep the contained air pure enough to oxygenate or arterialize the current of the circulation. The diffusion of the inflowing air throughout the air cells occasions a faint *respiratory murmur*, which may be heard by laying the ear on the chest. It is rougher or coarser over the trachea and larger bronchial tubes than over the pulmonary tissue. The sound in the one case is sometimes called *bronchial breathing*, in the other *vesicular breathing*.

.226. The inspired air, if fresh, contains only a trace of carbon dioxid, rarely amounting to .04 per cent., or 4 parts in 10,000 of the air. The expired air is warmer than the inspired air; it is saturated with moisture, as may be seen by breathing on some cold condensing surface; and it contains about 4.3 per cent. of carbon dioxid.

227. The act of respiration may be suspended voluntarily for a time; but the necessity for breathing speedily overcomes any voluntary restraint by spasmodic gasping efforts. When, notwithstanding these, aëration fails to take place, the blood tends to stagnate in the lungs, and is thrown back on the right ventricle, causing the veins to become distended and the surface congested and dusky. Meanwhile the circulation of venous blood in the brain benumbs the consciousness of the individual, whose struggles for breath become proportionately weaker. A person thus affected is said to be *asphyxiated*. When the lungs are sound and the deficient aëration of the blood is due to some mechanical interference with the inflow of air or some narcotic influence on the brain, the continuance of respiration by artificial means [417] is often of value in saving life. But difficulty of breathing is frequently due to causes which affect the integrity of the lungs or air passages, and which cannot be favorably affected by any method of artificial respiration. In *pneumonia* the air cells become filled

with exudation [173] from the capillary vessels, and the patient, in fatal cases, is to all intents and purposes drowned in the plasma of his own blood. In *diphtheria* the air tubes are narrowed and perhaps occluded by the formation of false membrane within them. In *bronchitis* the mucous membrane becomes swollen and blocks up the tubes, or an excess of secretion may produce the same effect. A slight inflammatory tumefaction of the lining membrane of the *larynx* has frequently caused death, for the air passage between the vocal cords [224] is so narrow as to be easily closed up. In *pleurisy*, or inflammation of the pleura, the pain of the inflamed surfaces rubbing against each other makes the breathing shallow, and later the lung may become compressed by an accumulation of effused serum in the pleural cavity. A *punctured wound of the chest* admits air into the cavity, and the lung becomes collapsed by the pressure of the atmosphere; during respiratory efforts air enters and escapes, as in the case of a leaky bellows, by the leak rather than by the regular passages.

228. When the same air has been breathed and rebreathed a number of times, it becomes unfit for respiratory use. Those who inhale it suffer from headache, heaviness of mind, and general discomfort. These effects are attributed rather to the organic emanations of the used air than to the carbon dioxid, for the latter, when derived from other sources than human respiration, may be inhaled in large proportions without harmful manifestations.

229. Fresh air contains about four volumes of carbon dioxid in 10,000; but as the average man throws out from his lungs .01 cubic foot per minute, or .6 cubic foot per hour, the quantity in an occupied room is speedily increased unless it be diluted and carried away by intermixture with a steady flow of fresh incoming air. The

air of bed-rooms, barrack-rooms, and hospital wards becomes oftentimes exceedingly foul at night. Instead of four volumes in 10,000, the air may be charged with twenty, forty, or more volumes. The quantity of carbon dioxid found in breathed air gives a definite expression to its vitiation. If half a fluidounce of lime water be introduced into a clean, dry, eight-ounce bottle containing the air to be examined, and shaken up vigorously for a minute or two, the appearance of a turbidity from insoluble carbonate of lime in the liquid will indicate the presence of eight or more volumes of carbon dioxid in 10,000 volumes of the air examined. If no turbidity be manifested, the air contains less than eight volumes in 10,000. Ready methods of air analysis, such as that just indicated, have not come into general use because they convey no more information than may be gathered by the sense of smell. A wholesome room should not have more than six parts per 10,000 of the air. When the carbon dioxid amounts to seven volumes, a want of freshness is observed on entering from the outer air; and when nine or more volumes are present, the organic odor becomes perceptible.

230. The inflow of fresh air to take the place of air vitiated by human occupancy is known as *ventilation*. Its object is to so dilute the products of respiration that the carbon dioxid of the air of the room shall not exceed six volumes in 10,000; and to effect this 3,000 cubic feet of external air are required hourly per man. An aperture of seventy-two square inches would furnish the needful supply if the inflow proceeded at the rate of two feet per second; but if the aperture were considerably smaller, ventilation could be effected only by the introduction of the air at such a rate as would cause danger from draughts. A corresponding aperture of exit must be provided.

231. By *natural ventilation* is understood that which takes place through doorways, windows, transoms, and the seams of imperfect woodwork, and, if there be an open fireplace, by the draught up the chimney. Heat is a powerful aid to ventilation. As warm air is lighter, bulk for bulk, than cold air, the warm air of a room rises to escape while colder air seeks to enter below. Even when there is no fire in the room the tendency of air vitiated by respiration is upward, because it comes from the lungs warmer than the surrounding air. When there are no special currents to deflect it, the puff of tobacco smoke from the lips of a smoker rises toward the ceiling. The natural exit for foul air is in the upper part of a room. To ventilate satisfactorily by windows, the sashes should be lowered from the top as well as raised from the bottom. But when ventilation is effected by artificial means, the air may be drawn off from any suitable point.

232. CUTANEOUS EXHALATION.—The body is enveloped in a tough, close-fitting elastic covering, continuous at the mouth, nostrils, and other apertures with the mucous membrane of the interior. This covering consists of the skin and an underlying layer of fatty tissue. Over the flexures of the joints, in the eyelids, and other parts the layer of fat is thin, that motion may not be impeded, while over the abdomen a thickness of several inches acts as a protection to the contained organs against injury from either violence or cold.

233. The skin consists of two layers: The *corium, derma* or true skin, which overlies the fat, is composed of closely interlaced elastic fibres, with vessels, nerves, hair follicles, small glands and channels for the passage of matters that are to be thrown out of the system. The upper layer, *cuticle* or *scarf skin*, consists of microscopic cells which

grow from the surface of the true skin, and are shed in a dried up or shrivelled condition from its own free surface. The cells of the cuticle are composed of an albuminous material, like the nails or hair, which are indeed outgrowths from its structure. Their horny nature may be observed on the palms of the hands after unusual labor, which by intermittent pressure has excited the cutaneous growth.

234. The skin is well supplied with nerves, which are distributed to small processes or *papillæ*, projecting from

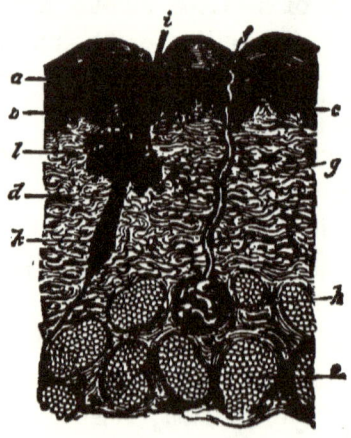

Perpendicular section of the skin, showing, *a*, the epidermis, cuticle, or scarf skin; *b*, a layer of dark-colored cells; *c*, the papillæ on the surface of *d*, the corium, derma, cutis vera, or true skin, and *e*, the fat cells underlying it; *f*, a perspiratory pore or aperture, *g*, the duct, and *h*, the coiled substance of a sudoriparous gland; *i*, the shaft of a hair, *k*, its root, and *l*, sebaceous glands communicating with the interior of the hair follicle.

the surface of the true skin like the villi [218] from the mucous membrane of the small intestine. These papillæ are particularly numerous on the ends of the fingers, where their arrangement in lines may be observed as parallel and sometimes concentric ridges. The sense of touch is proportioned to their size and number. The points of a compass, separated only one-twelfth of an inch, will be recognized as two distinct objects by the tips of the fingers; yet when separated two inches they will be felt to make but one impression on the middle of the forearm, thigh, or back. The cuticle gives protection to the sensitive papillæ by covering or sheathing them with an insensitive layer.

235. The skin is beset with hairs, each rooted on a small papilla at the bottom of a follicle. From this root it rises through the follicle to the surface and beyond. Small *sebaceous glands* are packed in the interstices of the true skin, their ducts opening usually into the hair follicles. They secrete an unctuous matter, to prevent the skin from being dried and cracked by exposure to the sun and air. On the face, and particularly on the nose, their ducts sometimes become choked and their apertures, soiled by exposure, appear as black points on the surface. Cautious pressure will liberate the retained secretion in the form of a tallowy cast of the duct.

236. But the elements in the structure of the skin by which the purification of the blood is mainly effected are the *sudoriparous glands*. These are small coiled tubules which reach from beneath the true skin to the surface of the cuticle, where their ducts terminate in minute pores. They are surrounded by a vascular network from which certain constituents of the blood are drawn off and transmitted to the surface to be dissipated as *insensible perspiration* or poured forth as a visible transudation. Condensed perspiration or *sweat* is a sour-smelling liquid, containing traces of saline and organic matter. The quantity of liquid eliminated daily from the blood by means of these glands depends on varying conditions as to exercise and external temperature; but its average is about two pounds. The

process is regulated by the nervous system; for in disorder of the latter perspirations may occur in the absence of the conditions which are ordinarily required for their production, as when the skin is pale and cold in the prostration of syncope [302] or approaching death.

237. Although the removal of water from the blood is one of the prime functions of the skin, the cuticle performs an important duty in preventing the general drying up of the tissues by evaporation. Moreover, when the water supply of the system is deficient, the skin may become an *absorbing* instead of an *eliminating* organ. Immersion in a bath of tepid water allays thirst. Even when the circulation is almost at a standstill, and the patient unconscious from a deficiency of water in the blood, he may be rallied from his dangerous condition by the absorbing power of the skin [405]. Nor is this power restricted to the absorption of water, for mercury may be taken up and the system brought under its influence by means of vapor baths or inunction. Absorption takes place readily from the surface of the true skin; and before hypodermic medication [502] came into use, the method of dusting or dressing a blistered surface with the medicine to be absorbed was occasionally practised.

238. URINARY EXCRETION.—The *kidneys* are two elongated organs situated on the wall of the abdominal cavity, one on each side of the vertebral column. The tissue of the kidney consists of the ramifications of blood-vessels in and around a series of fine tubes which drain the urine from them. These tubules unite on the inner aspect of each kidney into an elastic tube, the *ureter*, by which the liquid to be thrown out is conveyed to the bladder.

239. The *bladder* [161] is a strong muscular sac, coated in part with peritoneum and lined with mucous membrane.

Its discharge tube, the *urethra,* passes from its lower and fore part downward and forward beneath the pubic arch, and then along the under surface of the penis. In passing under the arch, the urethra penetrates a strong fibrous membrane which closes in the bottom of the pelvis; and this membrane it is which forms the impediment sometimes encountered in passing the catheter [294].

240. The *urine* consists of water holding in solution certain nitrogenous substances and inorganic salts. When the muscles are used up by exercise they are repaired by the albuminous principles of the blood, while their old and degenerated material is absorbed in the form of complex organic substances, which are finally converted into urea and filtered off through the kidneys. Some of the transition products in the decomposition of albuminous substances are occasionally found in traces in the urine; one of these, *uric acid,* is constantly present. On the other hand, in morbid states not only uric acid, but many of the less oxidized products of the decomposition of the albuminoids may be found in considerable quantities in the urine.

241. The inorganic salts of the urine, consisting of chlorids, sulphates, and phosphates of the alkalies and of lime and magnesia, are derived in great part from the food [198]. *Chlorids* are constant constituents of most articles of food. *Sulphates* originate in the decomposition of albuminous tissues, the sulphur of which becomes oxidized in the body. *Phosphates* exist naturally in both animal and vegetable foods; they are derived also from the reparation of the bony tissues and the oxidation of phosphorus in the disintegration of the nerves and nervous masses.

242. The *average quantity* of urine secreted daily is about fifteen hundred cubic centimeters, or fifty fluidounces. It is decreased in warm and dry weather, when cutaneous and pulmonary transpiration is active, for there is then less water left in the system for elimination by the kidneys; it is decreased also in diseases characterized by profuse watery evacuations from the bowels. It is increased in cold and damp weather, when exhalations from the skin and lungs are at a minimum, and in morbid conditions associated with dryness of skin, excepting those inflammatory or febrile diseases which tend to the suppression of all the secretions. Normal urine removes from the system daily about an ounce each of urea and salts and eight or ten grains of uric acid in the form of urates. When freshly passed it is slightly acid, but after a time becomes alkaline from the transformation of its urea into carbonate of ammonia.

243. ANIMAL HEAT.—When oxygen combines with carbon, carbon dioxid is produced, and heat is evolved during the process, which is called *oxidation* or *combustion*. The union of oxygen with carbon is followed by the same results, whether it takes place in the kitchen range or the human system. In the ordinary fireplace much of the heat that is produced passes off by the flue; so in the tissues much of the heat produced passes from them into the current of venous blood, and is thrown out by the lungs. Water takes a prominent part in the regulation of the animal heat. When a flame is applied to water in a flask, the heat is communicated to the water and its temperature rises until it reaches 212° F., when it boils; but after this its temperature rises no more, although the water still receives accessions of heat from the flame. Instead of becoming hotter it becomes converted into watery vapor or steam, and heat is absorbed during the process. Hence we can understand that every breath which sends out watery vapor into the atmosphere carries with it a certain amount of heat.

This is a cooling influence which is in constant operation to offset the continual heating influence of the oxidation which takes place in the tissues. The water evaporated from the skin is another cooling influence constantly in operation, but susceptible of augmentation or decrease in accordance with associated conditions. When, on account of exercise, an excess of heat is generated in the tissues, the breathing becomes deeper that more watery vapor may be thrown out at each expiration, and the perspiration is increased to cool the body by its evaporation. When, on the other hand, the vital actions are conducted so gently that there is no excess of heat generated, the water which would otherwise be exhaled from the skin is drawn off without loss of heat by the kidneys. These two organs supplement each other and keep the temperature of the body at a normal of 98.4° F. If, from any cause, as a chill to the surface, the perspiratory pores are closed, the heat of the body ceases to be dissipated and in a little while the condition which we call fever is developed. The influence of a diaphoretic, such as Dover's powder, in reducing fever by inducing perspiration, is thus understood. When a particular organ becomes inflamed, the active changes going on in its substance augment its heat, and a symptomatic fever [322] is established. When the action of the skin ceases because of a deficient supply of water in the system, the blood becomes heated and the circulation disturbed, leading to a condition which has been called *thermic fever*, but which is generally known as sunstroke [404].

CHAPTER III.

THE ADMINISTRATIVE SYSTEM.

244. The ADMINISTRATIVE SYSTEM consists of the brain, spinal cord, nerves, and ganglia. The *brain* is the nervous matter contained in the cranium; it is connected directly or intermediately with every organ and tissue as the *headquarters* of the body to which all the reports called *sensations* are rendered, and from which all orders for motion or action based on those reports are issued. The *spinal cord* extends downward through the vertebral canal, giving off branches or *nerves* at intervals from its sides. The *ganglia* are small fragments of nervous tissue found at intervals near the spinal column, or in protected situations, as behind the heart and around the large arteries of the abdomen. They communicate freely with each other by branches called *sympathetic nerves*, and with branches of the cerebro-spinal system; and their distribution is along the tracks of the arteries, going where the blood goes.

245. A large nerve consists of minute tubules enclosed in a strong protective sheath, as the insulated wires of a telegraph cable are enclosed in a non-conducting covering. The branching of a nerve is like the splitting up of the cable for the proper distribution of its contained wires. Each tubule, representing a wire, ultimately reaches its terminal station, where by its means messages may be conveyed to and from the brain.

246. The brain is a somewhat egg-shaped, pulpy body, divided from before backward into two contiguous *hemi-*

spheres; but the line of division does not extend all the way down, for about its middle there is a broad bridge of communication between the two sides. The surface presents also grooves called *sulci,* and as these give the brain the appearance of consisting of a packed mass of irregular coiled

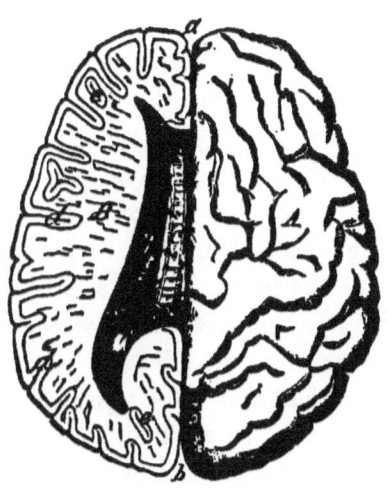

THE HEMISPHERES OF THE BRAIN: *A*, the right; *B*, the left, divided from before backward by *a, b,* the longitudinal fissure, and connected by *c,* the bridge of transverse fibres called the corpus callosum. On the right side the convolutions and sulci are shown; on the left the upper part of the convexity of the hemisphere has been cut away to show the gray matter *d, d,* dipping into the sulci and appearing as islands, *e, e,* in the interior of the white matter; the elongated cavity with curved extremities is the lateral ventricle of that side.

tubing, the seeming coils are spoken of as the cerebral *convolutions.* The exterior of the brain consists of *gray, cineritious,* or *vesicular* matter, the interior of *white* or *medullary* matter. The latter is a compact mass of nerve tubules connecting various parts of the gray matter with each other and with all parts of the body. There are several cavities, called *ventricles,* in the interior of the brain, their walls lined with gray matter. The brain weighs about fifty ounces, but the quantity of vesicular matter varies with the number of the convolutions and the depth of the sulci. The intelligence appears to be proportioned to the quantity of gray matter which envelops the white matter of the brain.

247. The interior of the cranium is lined with a strong fibrous membrane, the *dura mater*, which sustains and supports the brain, and supplies the nerves with a protective covering as they pass through the bony apertures. Internally it is faced with a serous layer, the *arachnoid*, which is reflected loosely over the convolutions, the two smooth surfaces lessening the risk of injury from jolting, just as the peritoneum [161] protects the abdominal organs from injury by friction. Beneath the arachnoid and in close contact with the brain matter is the *pia mater*, composed of a fine network of nutrient blood-vessels.

248. The *spinal cord* is protected by a strong sheath, which is lined like the dura mater. It consists chiefly of nerve tubules, but a curved streak of gray matter extends throughout its length on each side of its middle line. The nerves which emerge in the *cervical* region divide on each side into two sets, an upper and a lower. The former are distributed to the head, neck, and interior of the chest, one of them, the *phrenic*, being the medium by which the respiratory movement of the diaphragm [161] is sustained. The latter form an interlacement of large cords, called the *brachial plexus*, which crosses the upper part of the armpit with the axillary vessels, and is distributed to the upper extremity. The *dorsal* nerves supply the walls of the chest and abdomen. The *lumbar* and *sacral* nerves are distributed to the pelvis and hips, but several of them unite on each side into one large cord, the *great sciatic* nerve, which extends along the back of the thigh, and breaks up into branches for the supply of the lower extremity.

249. There are two kinds of nerve tubules: *Nerves of sensation*, or *sensory nerves*, convey information to the gray matter of the brain; *nerves of motion*, or *motor nerves*, transmit from the gray matter the influence by which the muscles are brought into action. When a sensory nerve is cut across, the area of its distribution loses its feeling because the information concerning its condition does not reach the gray matter of the brain. When a motor nerve is severed, the muscles which it supplied become paralyzed, because the orders emanating from the gray matter do not reach them. Most of the nerves contain both motor and sensory filaments; but some are purely motor, as the *facial*, a cerebral nerve, which supplies the muscles of the head and face. On its way out of the cranium it traverses the middle ear [259], and when it becomes involved in certain destructive diseases of that organ, the corresponding side of the face loses the power of motion, becoming blank and expressionless, but retains sensation, for the sensory filaments are derived from another nerve. The motor filaments of the spinal cord are aggregated in its anterior part; the sensory filaments in its posterior part. Hence the lower limbs of a patient may be paralyzed and yet retain their sensation, because the posterior part of the cord has not suffered from the disease or injury so much as the anterior part.

250. The nerves from certain organs bring special information to the brain. The optic nerves transmit impressions as to light or shade, color, form, size, distance; the auditory nerves bring impressions of sound; those of taste and smell guard the entrance to the alimentary and respiratory tracts; and from certain parts, as the tips of the fingers, come sensations giving definite ideas as to certain properties of external objects.

251. The *eye* is protected by its deep situation in the socket, by the projecting ridge of the frontal bone and the quick movement of its upper lid. It is a small, globular-shaped camera, covered in front by mucous membrane, the *conjunctiva*. The rays of light enter through the *cornea*, which is set into the front of the eyeball like a watch-glass

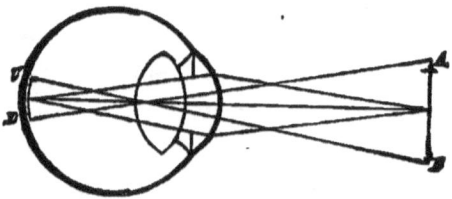

The rays from the object *A B* are brought to a focus on the retina at *C D*, where an inverted image is formed as in the camera of the photographer.

over the dial plate. Behind the cornea is a colored curtain of muscular fibres, the *iris*, with a central aperture, the *pupil*, which becomes contracted or dilated, according as the light is strong or weak. The *crystalline lens*, situated behind the iris, brings the rays to a focus on the *retina*, a delicate expansion of the *optic* nerve. The whole is enclosed, except on the corneal front, in a strong casing of white fibrous tissue, the *sclerotic* coat.

252. When a camera is out of focus the lens is moved forward or backward by a screw until the image is clearly outlined on the ground-glass screen. In the eye a small muscle increases the refracting powers to bring the rays from near objects to a focus on the retina. This is called the *power of accommodation* of the eye.

253. An individual should see distinctly at six hundred yards a black centre, three feet in diameter, on a white ground, or at twenty feet a circular spot, four-tenths of an inch in diameter, on a white card. An eye which responds to this test is *emmetropic* or normal.

254. In elderly people vision for distant objects remains normal, but that for near objects is impaired. In reading, the sight has to be aided by convex glasses. This condition is called *presbyopia*.

255. When the lens is too convex or the refracting media of the eye too powerful, the rays from distant objects are brought to a focus *before* they reach the retina. The spot on the card must be brought nearer than twenty feet to be clearly distinguished, or a concave glass must be placed before the eye to diminish its refracting power. This is the condition of *myopia*, or *short-sightedness*.

256. When the lens is too flat or the refracting media too weak, the parallel rays from a distant object fail to reach a focus until they have passed *behind* the retina; and bringing the object nearer increases the difficulty. This, the condition of *hypermetropia* or *long-sightedness*, requires convex glasses to aid the eye in focussing the rays.

257. When the cornea is irregular in its curvature, a vertical line may be seen clearly, while a horizontal line crossing it is blurred or indistinct. This defect, *astigmatism*, cannot be remedied if the corneal surface is so faulty as to distort the outlines of objects like a piece of poorly made window-glass; but if there is merely a bulge or a flattening at some particular point, special glasses may be fitted.

258. *Color-blindness* is an inability to recognize certain colors. When a man who is *green-blind* is presented with a skein of worsted of that color and desired to pick from a stock of various colors the skeins which correspond with it in color, irrespective of shade, he will select and add to his

green samples skeins of other colors, mostly neutral tints, as grays, drabs, etc., which seem to him to have the same depth of shade as his test skein. When he is presented with a roseate skein—one having a tinge of purple in its redness—he will match it with shades of green or gray; and when tested with a bright-red skein he will bring forward light greens and browns as corresponding with it in color. A *red-blind* man can see the purple but not the red in the roseate test skein, and so selects light shades of blue and violet as samples of the roseate color. When presented with a bright-red skein, he matches it with shades of green and brown that seem to the normal eye to be darker than the test skein.

259. The *external ear* consists of a framework of pliant cartilage bound to the bony parts by strong fibrous tissue. At the bottom of the shell, or *concha*, is the opening of the *external auditory canal*, which conveys the vibrations of the air to the *tympanum* or *drum* of the ear. Its entrance is beset with hairs to exclude insects, and its sebaceous follicles [235] secrete a bitter waxy substance. This canal extends inward and forward for about an inch, curving slightly, so that its middle part is somewhat higher than its ends [430]. The *middle ear* consists of a small cavity in the interior of the temporal bone. It is lined with a delicate mucous membrane, and any mucous secretion that may gather in it drains off by a tubular passage, the *Eustachian tube*, into the back part of the throat. Ordinarily the walls of this tube are in apposition, but during swallowing they become separated. In blowing the nose, air is sometimes forced into the tympanic cavity, giving rise to a feeling of distention in the middle ear, which is, however, immediately relieved by the movement of swallowing. When the Eustachian tube is obstructed, as frequently happens from the swelling of an ordinary sore throat, a temporary deafness may be developed because air cannot get into the tympanic cavity to support the drum of the ear on its inner side against the pressure of the external air on its outer side. The *internal ear* is a complex arrangement of the nervous tissue by which the vibrations of the tympanum are conveyed to the brain.

260. The *olfactory nerve* is distributed to the mucous membrane covering the upper part of the interior of the nose. It enters from the brain through apertures in a thin plate of bone which forms part of the floor of the cranium and the roof of the nasal cavity.

261. The *tongue* is the organ of taste, but some special nervous filaments are distributed to the palate and fauces. It is supplied also with nerves of common sensibility, which enable it to detect extraneous matters in the food during mastication. It consists of longitudinal and transverse muscular fibres, which by their contraction enable it to assume almost any form. Its surface is roughened with close-set papillæ which are covered with mucous membrane. A longitudinal furrow on its upper surface ends behind in a depression near the base of the tongue; and from this point some large papillæ project outward and forward, in the form of a V. To impress the nerves of taste a substance must be in solution or soluble in the liquids of the mouth. If the substance is insoluble it is recognized only by the nerves of common sensation as insipid, gritty, etc.

262. In the government of the army only such matters as affect the general welfare are forwarded for the action or decision of the War Department; minor and routine matters are disposed of by subordinate commanders. So in the human system, many sensations or reports are trans-

mitted by the nerves, which are not of sufficient importance to call for special action by the INTELLIGENCE at headquarters; these are disposed of by what is called *reflex action*. When the toes of a patient are tickled or pinched, the muscles of his thigh and leg are brought into action, and the foot is drawn away. This may be a voluntary act. The sensation may be received and appreciated by the brain, and in consequence a motor influence may be sent out to effect the withdrawal of the foot from the cause of the irritation. But, on the other hand, it may be involuntary and independent of sensation, as when it occurs in a paralyzed limb. In such a case the gray matter of the *spinal cord* takes note of the sensation, recognizes it as being not of sufficient importance to be forwarded to the brain, and itself sends out orders to meet the needs of the case. Most of the routine acts of life are regulated by this action of the gray matter of the spinal cord. When an individual walks, the pressure of his weight on the sole of the foot is transmitted to the cord, and elicits the motor influence for the continuance of the movement. Separate acts of volition are not required for the various motions that make up each step of a prolonged walk. But it is in the regulation of the functions of the various organs that reflex action has its greatest field of operation. The whole of the business of organic life is conducted by the nervous matter of the ganlia [244], and the gray matter of the brain receives intelligence only when something goes wrong. When the ventricles of the heart become filled with blood the sensory filaments report the fact to the cardiac ganglion, and orders are issued for immediate contraction. When there is an insufficiency of oxygen in the lungs the ganglionic nervous matter is informed, and, according to the necessities of the case, a deep or shallow inspiration is taken [223]. When masticated food reaches the fauces the muscles of deglutition are called into action. When food enters the stomach its presence is announced and, from a special ganglion, the standing orders for the needful increase in the supply of blood and the prompt secretion of the gastric juice [211] are carried out. Contact of the chyme with the intestinal lining produces the vermicular or peristaltic movements [216], which carry the alimentary mass onward from coil to coil. The functions of the kidneys, liver, skin, etc., are all thus regulated and co-ordinated by reflex influence. This system is sometimes called the sympathetic nervous system, because by means of its connection with the cerebro-spinal nervous filaments the organs which it regulates are to a certain extent brought into relation or sympathy with the mind or will. The cheeks, for instance, flush or pale, and the heart palpitates or fails under the influence of mental emotions, and the will exercises a limited control over the respiratory movements.

263. Reflex action from ganglionic or spinal nervous matter relieves the gray substance of the brain from the care of supervising the normal action of the organs and the routine movements of the body, leaving its energies to be expended in the exercise of the mind and will. That the INTELLIGENCE finds enough to do in this way is shown by the necessity for sleep. The brain is well supplied with blood to repair its waste, part of which appears in the urine as phosphates [241]; yet the rapidity of the waste is such that a certain number of hours in every twenty-four must be given up wholly for recuperation. Nothing manifests the independent character of the reflex acts so clearly as their continuance during sleep, when the intelligence is profoundly suspended. The heart beats, the chest heaves, the arterialized blood circulates, the liver, kidneys, skin, etc.,

perform their various functions under the controlling influence of the ganglionic system, and if the position of the body or limbs be in any way uncomfortable, the gray matter of the spinal cord intercepts the message, and rectifies the position without disturbing the sleeping INTELLIGENCE.

PART III.

THE SPECIAL DUTIES OF THE HOSPITAL CORPS.

CHAPTER I.

MANAGEMENT OF ACCIDENTS, ETC.

264. A seriously wounded or sick man must be removed to hospital without delay, except when immediate treatment offers the only hope of saving life.

If the patient be insensible, place him on his back with a support under his head; remove his neckwear and unbutton the clothing on his chest. Then sprinkle cold water on his head, face, and chest, and put some to his lips to ascertain if he has the power of swallowing. The nature of the case should be made out, with a view to relieving urgent symptoms and preventing injury during transportation.

265. If the injury be obvious, ascertain its characters; if obscure, inquire of the bystanders what is known concerning its history or cause. Personal examination of the patient must be made with the utmost gentleness, lest harm be done to some undiscovered injury. When facilities are available, the case should be treated as at the first dressing-station of field service [41, 365]. All symptoms observed during this examination should be noted mentally for the information of the medical officer, because some change may take place in the condition of the patient before he comes under medical supervision.

266. When there is much prostration, the patient's head must be kept low; and if he has lost so much blood as to be notably weak or faint, efforts should be made to revive him before removing him to hospital. In no case while in this exhausted state should he be placed on his feet or required to make exertion, however slight, in his own behalf lest he fall into the condition of syncope. If the case is one of external hemorrhage, an ounce of brandy or whiskey, with .020 morphine, should be administered before placing him on the litter; but if the hemorrhage is internal, faintness should be relieved by small doses, twenty or thirty drops, of aromatic spirit of ammonia at short intervals in a spoonful of water.

267. In moving a case of fractured thigh or leg, one person should support the fractured parts; and before transporting the patient on the litter the foot of the injured limb should be tied to the other, to prevent its rolling outward by its own weight; and support, as by a blanket rolled into a cylinder, should be applied along the outer side of the limb.

268. At the hospital the patient is prepared for bed by removing his clothes, ripping the seams, if necessary, cleansing soiled surfaces, and applying dressings. When the patient is unconscious or paralyzed, the mattress should be protected by rubber or gutta-percha cloth under the sheets, and a folded blanket, covered with a draw-sheet, should be placed under the hips.

269. The private of the hospital corps on ward duty should be an intelligent, careful, and even-tempered man. His duties as relating to the routine work of the ward are easily learned; but a long experience is needful to fit him for those pertaining to the treatment of the sick. He is responsible for the ventilation, heating, cleanliness, and discipline of his ward. He stands by the bedside during the visit and listens attentively to the instructions of the surgeon. He obtains the medicines from the dispensary, and administers them as directed; and in special cases feeds the patients and attends to all their personal wants. These duties require tact and gentleness, for sick men are often irritable and difficult to please. He should note carefully all changes in the symptoms for the information of the steward and medical officer. Changes in the pulse, temperature, and respiration are recorded in most of the serious cases; the frequency and character of the discharges, the occurrence of cerebral disturbance, the times and duration of sleep, and any complaint of pain or any abnormal sensation felt by the patient should also be carefully recorded. The more a nurse knows of the condition of a patient the greater is his value, for inexperience and want of knowledge will often call unnecessarily for assistance, while, on the other hand, they may consider as unimportant some change which is of vital interest.

270. Ordinarily the patient should have a bath before being put to bed. A *warm* bath, from 85° to 98° F., soothes the system, predisposing to sleep, while it cleans the skin and promotes its activity. A *tepid* bath, from 65° to 85°, has little value in hospital practice. A *cold* bath, below 65°, and a *hot* bath, above 98°, are powerful influences: the former, when the immersion is short, is followed by warmth and tingling of the surface; the latter quickens the pulse and respiration, congests the surface, and may induce faintness. When the hot bath is followed by free perspiration its action is salutary; but when it occasions fulness in the head, throbbing, and giddiness, its influence is

harmful. Cold baths are seldom used for the sick, as a certain vigor of constitution is needful to enable reaction [304] to take place. The *hot-air bath* is often conveniently substituted for the hot bath, as it may be given by the bedside, the patient sitting on a cane-bottomed chair, with a spirit-lamp beneath, and blankets draped around him from the neck to the floor to confine the heated air.

272. Preparations for the performance of a surgical operation can be made only by one who has a knowledge of what is to be done. Preliminary to operation the surgical staff and nurses, the dressings, instruments, and patient, must undergo a systematic preparation.

Disinfection of the Hands.—All rings are removed, and the hands and forearms are scrubbed for five minutes with soft soap and hot water, particular attention being given to the nails; after drying, the parts are washed with alcohol and then in a solution of bichlorid, 1 to 2,000, in warm boiled water. Some surgeons, after washing with soft soap, make the skin mahogany brown with permanganate of potassium, subsequently decolorizing with oxalic acid solution, and then lessening the irritation of the skin by washing in boiled lime-water.

Disinfection of Dressings.—These are steamed in a sterilizer for twenty minutes, after which they are removed with aseptic hands to jars which have been scrubbed with soft soap and afterward with bichlorid solution. The contents of the jars are covered with sterilized towels soaked in bichlorid.

Disinfection of Instruments.—Every knife or sharp-edged instrument is wrapped in gauze, to preserve its edge from injury, and boiled for five minutes in a one-per-cent. soda solution. Each is then removed with aseptic fingers, and all are arranged in pans containing sterilized water and covered with sterilized towels.

The needle of a *hypodermic syringe* is boiled with the instruments; but the barrel and piston are soaked an hour in carbolic acid, five per cent., in alcohol.

Silk thread or *silver wire* is wound on glass rods or spools, and boiled with the instruments or steamed with the dressings. *Rubber drainage-tubes* and *catheters* are boiled or steamed; *silkworm gut* is boiled five minutes in alcohol; and catgut is soaked in ether for two days and then put in alcohol in a firmly corked bottle, which is boiled in water for two hours.

Disinfection of the Patient.—If possible, on the day before the operation the patient is given a bath; the area of operation is shaved, washed with soft soap and water, rinsed off with boiled water, washed with alcohol, and scrubbed with bichlorid solution, after which the part is covered with aseptic gauze, secured with a bandage. On the day of the operation the patient is bathed and given a clean flannel shirt. The area of operation is then treated as on the previous day, and the patient placed on the operating-table and covered with a sterilized sheet. All persons assisting should wear sterilized linen gowns, the hair covered with a sterilized towel, and the hands disinfected; and whenever an object not aseptic has been touched, the hands should be bathed in bichlorid again.

273. A table of suitable height and length should be covered with two or three folds of blanket, overlaid by a disinfected rubber cloth, and provided with a low pillow for the head. Chloroform or ether [276] is then administered. The limb, if an amputation is intended, is then elevated and emptied of blood by pressure with the hands in the direction of the trunk, after which the tourniquet [387], which has already been placed in position, is screwed up to stop the circulation. Esmarch's bandage [275] may

be used to empty the limb, and his hollow rubber cord to control the inflow of blood. The surgeon now cuts the soft parts to the bone, and the flaps, or incised tissues, are in-

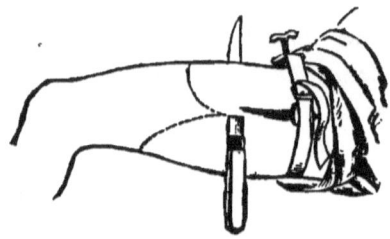

Amputation of the thigh.

stantly gathered up in the hands of an assistant, who draws them in the direction of the trunk, out of the way of the saw. Sometimes, instead of his fingers, the assistant uses

A bone-cutting forceps.

for this purpose a *retractor*, a piece of muslin two feet long, and as wide as the diameter of the limb operated on. This is torn for half its length down its middle, and the

A spring or artery forceps.

two strips are made to embrace the bone at their point of junction, so that by traction on the muslin the soft parts are protected. During the application of the saw the assistant in charge of the limb is careful to do no more than support its weight; if he do more than this he will lock the saw between the freshly sawn surfaces, and if he do less the unsupported part of the weight may splinter off the under part of its upper fragment. On the removal of the limb the surgeon smoothes the edges of the bony stump with a scraper or strong-backed, blunt scalpel, and may have to use a bone-cutting forceps to round off sharp angles. The prominent arteries are then picked up by forceps or tenaculum, and tied [389]; and the tourniquet is gradually relaxed to permit those that remain unligatured to indicate their position by their leakage. When these are secured, the tourniquet is removed and the cut surfaces are well washed with sterilized water until all oozing has ceased, after which they are carefully bathed with an antiseptic lotion, united by sutures [356], and dressed.

274. The provision for such an operation includes, therefore, a suitable operating-table; an anæsthetic, and handkerchief, or cone and sponge, for its administration; hot and cold water supplies, basins, gauze, and towels; disinfecting solutions; amputating knives, among them a double-edged

A tenaculum.   A scalpel.

one, or *catlin*, if the forearm or leg be concerned in the operation; scalpels, for incidental use, and one with a strong back as a bone-scraper; a bone-cutting forceps; several artery forceps, tenacula, and ligatures, with needles and prepared silk or gut for sutures; dressings, and brandy or whiskey for use in case of need.

275. Esmarch's bandage is a broad elastic roller, applied from the toes or fingers upward, to compress the soft parts and drive the blood before it toward the trunk. When the bandage has reached a point three or four inches above the

site of the intended operation, a piece of strong elastic tubing or flat band, with hook and chain attachment, is wound several times around the limb to cut off its circulation, after which the bandage is removed and the limb is ready for operation. The removal of the strong tubing, which acts the part of tourniquet in this method, is often followed by profuse oozing. To prevent it, some operators close the wound and elevate the stump before entirely removing the tubing; others attempt to control it by the application of hot water, 150° to 180° F.; but perhaps the best method is that which shortens the period of pressure by the substitution of the ordinary tourniquet for the tubing as soon as the principal arteries have been tied.

276. *Anæsthetics* produce insensibility to pain. *Chloroform, ether,* and *nitrous oxid* or *laughing-gas* are the anæsthetics chiefly employed. The last-mentioned is used by dentists in the extraction of teeth; the evanescent nature of its effects renders it unsuitable for operations that require time for their performance. The two others are used in general surgery. They induce a condition of excitement which is followed by insensibility. After a few inhalations the face becomes flushed, and the pulse and breathing are excited; the patient may be talkative and noisy, jovial or quarrelsome, but is always incoherent, and often struggling in his delirious fancies. Afterward muscular power and sensation are lost, as evidenced by the inability of the patient to sustain the weight of his arm when it has been raised by the surgeon, and his failure to wink when the conjunctiva is touched. His breathing is quiet, and his pulse full and regular, but perhaps somewhat slower than natural. The object of the administrator is to prolong this condition to the close of the operation; if he give too little, the patient may begin to struggle; if he give too much, stupor with stertorous breathing is developed. Usually the movements incident to dressing the wounds arouse the patient; failing this, consciousness is recalled by slapping the forehead and chest lightly with a cold, wet towel.

277. A larger quantity of ether than of chloroform is required to produce the anæsthetic effect; its vapor is more irritating, causing cough, feelings of suffocation, and disinclination on the part of the patient to continue the inhalation; it is highly inflammable, and hence dangerous under certain conditions; the excitement produced is more violent, and its period more protracted, particularly if the vapor be much diluted. The respiration under ether is less regular, being often shallow and spasmodic. Its after-effects—headache, confusion of mind, nausea, and general malaise—are of greater intensity, and last longer than those which follow the use of chloroform. The latter, therefore, has many points in its favor; but these are counterbalanced by the greater risk of death: many more deaths from chloroform than from ether have been reported.

278. Chloroform is best administered from a conical sponge, or a soft handkerchief folded to a square of somewhat more than four inches. One or two drachms are poured on, and the patient is told to breathe deeply, while the sponge or handkerchief is held over his mouth and nostrils at such a distance as will permit a mixture of atmospheric air with the indrawn current of anæsthetic vapor. As soon as the handkerchief loses the freshness of the chloroform odor, another similar quantity is sprinkled over it. As the vapor is much heavier than air, there is no need of oiled silk or other impervious covering to the upper surface of the sponge or handkerchief. During the period of excitement, the quantity of entering air should be diminished; but as soon as the patient sinks, with a deep breath or sigh, into a condition similar to that of quiet sleep, the handkerchief should be raised to admit of a freer entrance of air. The free admission of air is important; a small quantity of chloroform vapor permeating the system to the exclusion of air is far more dangerous than a larger quantity administered with proper precautions. If the patient show any sign of returning consciousness, a freshening of the chloroform supply, or a lowering of the handkerchief to increase the percentage of vapor inhaled, is required; and, on the other hand, if the breathing begin to be stertorous, the handkerchief may be laid aside for the time being. But meanwhile the administrator must permit nothing to interfere with his guard over the pulse and respiration of the patient. The slightest irregularity in either dictates the temporary withdrawal of the vapor. When chloroform proves dangerous, it is usually by a stoppage of the heart, sometimes of the breathing. When this happens, the operation must be suspended, fresh air admitted into the room, the patient placed with his head considerably lower than his body, his tongue pulled forward, and breathing encouraged by slapping the chest lightly with a cloth wet with cold water, or artificial respiration must be employed. The danger of vomiting is lessened by the administration of thirty cubic centimetres of whiskey before beginning the inhalation. In case vomiting occurs, the patient should be rolled over on his side, that the vomited matters may have free exit from his mouth. Death may occur from suffocation if these matters find their way into the trachea. If the stomach of the patient is comparatively empty at the time of the administration, the risk attending vomiting is materially lessened.

279. Ether is administered from a sponge covered on its upper surface with a layer of oiled silk. Sometimes a folded

towel is rolled into the shape of a hollow cone, open at the base for application over the mouth and nostrils; the ether is sprinkled over its interior. A cone of stout packing-paper is also occasionally used, with a sponge fastened in its apex. A half ounce or more is poured on the sponge at a time, and the patient watched as in the case of chloroform. Danger in the case of ether usually arises from failure of the respiration; less frequently from stoppage of the heart.

280. *Local anæsthesia* for subduing the pain in minor operations is produced by throwing a spray of ether on the part. The cold resulting from the evaporation of the atomized ether benumbs the skin, and, if not too prolonged, is not followed by undesirable results.

281. The hydrochlorate of cocaine, in a four or five per cent. solution, is used in operations on mucous surfaces. It is often used in operations on the eye. A few drops paralyze the terminal branches of the nerves of sensation, the effect lasting only five or ten minutes. It is used also in operations on the mouth and throat, rectum, vagina, and urethra; twenty-four grains injected into the rectum have proved fatal. It has no value as an anæsthetic when applied to the skin.

282. NECESSITY FOR CLEANLINESS IN THE TREATMENT OF WOUNDS, OPEN SORES, ETC.—The discharges from wounds, ulcerations, abscesses, etc., have always a putrefactive or septic tendency; and this should be constantly borne in mind by those who are concerned in the management of surgical cases, that the most perfect cleanliness, combined with antiseptic precautions and free ventilation, may effectually prevent one wound from being injured by the less favorable conditions of another. Hospital attendants have, indeed, to make perfect cleanliness their guiding principle even for their own welfare, for the discharges

from mucous surfaces and cutaneous or other ulcerations have, in certain diseases, a contagious quality. Gonorrheal matter coming in contact with the eyes, as from a tainted towel or finger, will light up a most violent and destructive inflammation [425], and a hangnail or other trifling abrasion on the hand may permit the introduction of the syphilitic virus.

283. The *diet of invalids* is based on our knowledge of the action of the stomach on food. As starchy and saccharine articles, when properly prepared in a semi-liquid condition, do not make any call on the powers of the stomach for their digestion [211], they are largely used in the sick chamber. Arrowroot, barley, corn-starch, farina, and tapioca are invariably found in the list of hospital stores. Nitrogenous substances, when used for the sick, are reduced to the liquid form, as beef tea, mutton, or chicken broth, etc. The patient takes wine whey until he is able to digest the casein of milk, and eggs beaten up with wine or brandy until the stomach can digest more solid forms of nutriment. Jellies lead on to custards, light puddings, and various farinaceous dishes, and these to fish, chicken, and carefully cooked meats.

284. The appearance of the tongue often gives important information concerning the condition of the digestive organs and the general system. Naturally the tongue is clean, red, and pliant. In malarial disease it is large, flabby, coated with a bluish or grayish film, and indented on the margin as if with the impression of the teeth. In disordered digestion it is cream-coated or covered with a yellowish fur. When notable fever is present it becomes dry; a moist tongue is always an improvement on a dry one. In typhoid fever it is narrow, elongated, dry, with a dark stain on the centre of its dorsum, and of a deep red color at its tip and edges. Sometimes it cracks and bleeds, and the blood dries into dark crusts, called *sordes*, on the tongue, lips and teeth. Medicines that are unpleasant to the taste may be deprived of much of their flavor by closing the nostrils while swallowing them.

285. Free vomiting is less distressing than dry retching. A patient suffering from the latter condition should be liberally supplied with tepid water or demulcent drinks, the more so as their subsequent ejection facilitates the removal of obnoxious matters from the stomach.

286. The natural temperature in man is usually said to be 98.4° F., but it varies a little, being more frequently above than below that point. When it falls below 97°, the patient is in a state of collapse [303]; when it rises above 99.5°, he is feverish. Up to 102° the fever is said to be moderate; above that degree it is high. Death is imminent when the temperature reaches 106°.

287. Bodily temperatures are sometimes recorded in degrees of the Centigrade scale. The zero or freezing-point of this scale corresponds with 32° F.; and the boiling-point, 212°, of the latter with 100° Centigrade. Hence 180 Fahrenheit degrees $= 100$ Centigrade, or $1 = \frac{5}{9}$. To change Centigrade statements into the Fahrenheit scale, multiply by $\frac{9}{5}$ and add 32. To change Fahrenheit into Centigrade, subtract 32 and multiply the remainder by $\frac{5}{9}$.

288. The *clinical thermometer* is a mercurial column, self-registering by a detached portion of the mercury, and accurately graduated for the short range of temperature likely to be observed above or below the normal. A sudden jar communicated to the hand which holds the instrument, bulb downward, will send the index down when required. When the patient is conscious and intelligent, the temperature is taken by placing the bulb of the thermometer under the tongue, and having him close his lips around the stem and breathe gently through his nose for two or three minutes, or until the mercury ceases to rise. The temperature may also be taken by placing the bulb in the armpit and pressing the arm close to the chest during the time of the observation. The rectum is a good situation for obtaining the temperature in unconscious patients.

289. The normal temperature is subject to a daily fluctuation of about a degree of Fahrenheit's scale. It is high-

Temperature chart, showing the morning and evening observations in a case of malarial fever; quinine was given at the times marked *, and its influence is manifested in the succeeding fall of the temperature.

est in the evening, at from five to eight o'clock; lowest during the night or early morning, from two to six o'clock. The hours of low temperature enhance the danger in cases of low vitality. When fever is present, these fluctuations are often distinctly marked. In the evening the patient becomes manifestly worse; the flush deepens on his cheek, the skin becomes hotter, and the cerebral disturbance aggravated. These are called *evening exacerbations*. The subsidence of this excess of fever after midnight is spoken of as a *remission*. Evening exacerbations occur in all fevers, even in those which, like typhoid, are regarded as continued fevers. When the remission is strongly marked or occurs daily at an unusual hour, the fever is a *remittent;* and when, during the lull, the temperature drops to the normal or below it, the fever is said to *intermit.* The taking of the temperature in febrile cases is an important duty. It not only shows the progress of the fever and the influence of quinine or other anti-febrile remedies upon it, but often enables the physician to prognosticate the result. A line drawn across a ruled scale so as to show the morning and evening temperature daily during the course of the disease is called a *temperature chart.*

290. When the lower limbs are paralyzed, the condition is called *paraplegia;* this is occasioned by disease or injury of the lower part of the spinal cord. Fracture or displacement of the vertebræ may have severed the cord or caused pressure upon it, or the pressure may be the result of inflammatory effusion or exudation [173]. If the lesion be somewhat higher in the loins, there will be involuntary passage of urine and fæces. When the lesion is high in the dorsal region, respiration becomes oppressed from paralysis of the muscles concerned in expanding and contracting the chest. When in the lower cervical region, the upper extremities participate in the paralysis. When above the origin of the phrenic nerves, death follows immediately from paralysis of the respiration.

291. When paralysis is due to disease or injury of the brain, as concussion or compression [401], apoplexy [403], etc., the loss of motion and sensation is one-sided, *hemiplegia;* but as the respiratory nerves of the unaffected side carry on the vital movements, death does not follow so promptly as when the upper part of the cord is involved. In hemiplegic cases the paralysis of the limbs is on the side

opposite to that which is diseased or injured, for the nerves from each side of the brain cross over to the opposite side as they leave the cranium to enter the spinal canal; but as the nerves which supply the head and face are cerebral nerves issuing above this crossing of the tubules, the side of the face which is paralyzed is that in which the lesion exists. An apoplectic clot in the right side of the brain paralyzes the right side of the face and the left side of the body.

292. To change a soiled sheet under a paralyzed or helpless patient: Turn him on his side; roll the uncovered half of the soiled sheet up lengthwise until the cylinder lies close to his back. Have one-half of the fresh sheet rolled into a similar roll and place its cylinder alongside of the other, smoothing the unrolled half over the bed and tucking it away properly. Turn the patient on his other side so as to bring him on the clean sheet; remove the soiled sheet, and unroll and smooth out the fresh one.

293. *Bed-sores* are usually found in patients who have been confined to bed for a long period by exhausting diseases, paralyses, compound fractures, etc.; and their site is the tissues subjected to pressure between the mattress and some bone that is not well padded with soft parts, as the sacrum and the prominences of the hip-bones, spine, or ribs. The treatment consists of regularly inspecting the back and other points of possible injury in a patient likely to become affected in this way, and frequently changing his position to relieve the threatened parts from continued pressure. The skin should be kept dry and clean by washing with soap and water, or bathing with alcohol, or a solution of corrosive sublimate, 1 to 1,000, and afterward dusting with fine starch. The sheet on which the patient lies should be kept free from creases and crumbs, and be replaced by

fresh linen as soon as it becomes soiled or damp. Benefit may also be derived from the protection of a piece of soft leather, spread with soap plaster, and the judicious use of pillows. In unconscious or paralyzed patients, the skin should be kept free from the irritating contact of discharges from the bowels or bladder; but when this is impossible, it should be protected by simple cerate or petrolatum. When a sore has formed it should be relieved from all pressure by air cushions or other suitable rings or pads, poulticed if sloughing, and antiseptically dressed if discharging only molecularly. When paralysis is due to fracture of the spine [449], bed-sores may be looked for in the course of a few days, as in addition to the pressure there is a loss of preservative nerve power in the parts. Such cases require every care in nursing to prevent the sores from themselves becoming the active agent in causing the death of the patient.

294. For the *introduction of the catheter* the patient should be on his back, head and shoulders slightly raised by pillows, knees drawn up and separated from each other to relax the muscles of the abdomen. The operator stands on the left side of the bed, and, supporting the penis with the fingers of his left hand, inserts the point of the instrument, previously warmed and oiled, into the mouth of the urethra. In doing this the body of the instrument is almost horizontal, its concavity looking to the left groin. As the point passes down the urethra the handle is swung round horizontally from the groin to the middle of the abdomen and is then gradually raised to the perpendicular and depressed between the thighs of the patient as the urine begins to flow through it. A suitable vessel should be placed to receive the discharge and prevent unnecessary soiling of the bedding. A piece of rubber tubing fitted to

the mouth of the catheter and leading into the vessel is sometimes useful. No force should be used in the operation, the catheter being guided rather than pushed into the bladder. Sometimes an impediment is offered by the point of the instrument to the movement which brings the handle to the perpendicular. This is best overcome by withdrawing the instrument for an inch and letting it glide farther into the canal before attempting to raise the handle. A large-sized catheter should always be employed for catheterization unconnected with a strictured urethra.

295. When the urine is passed in large quantity [242] it is usually pale in *color;* when scanty, high-colored. Blood, when present, may be identified by the appearance of the red corpuscles under the microscope. When in small quantity and diffused throughout the urine, it probably comes from the kidneys; when in larger quantity or in clots, and when it escapes more freely during straining at the close of urination, it probably comes from the ureters or bladder; when it comes before the flow of urine, which afterward becomes clear, its source is probably the lining of the urethra. In some acute malarial attacks the urine is of the color of blood, although no red corpuscles can be discovered by the microscope. Some substances, as rhubarb, beetroot, logwood, etc., may give rise to an appearance as of blood in the urine.

296. The *specific gravity* of urine is usually said to be 1.020; but as this depends on the quantity of dissolved solids that pass off in the liquid, it necessarily varies according as the urine examined was secreted after fasting or after the ingestion of food or drink. In the absence of an accurate urinometer the specific gravity may be taken by balancing a small glass vessel on the dispensing-scales and delivering into it from a pipette ten or more cubic centimetres of the urine. The weight of the urine divided by the number of cubic centimetres indicates the specific gravity. If the contents of a 25 c.c. pipette weigh 25.500, the specific gravity is $25.500 \div 25 = 1.020$. A notably low specific gravity is suggestive of *albuminuria* or *Bright's disease;* a high specific gravity, of *diabetes,* a disease in which sugar is present.

297. When urine is bloody it necessarily contains *albumin;* but this substance in certain morbid states escapes from the blood into the secreting tubules of the kidneys along with the urea and salts which constitute the natural excretion. Its presence in urine is suggested by the persistence of air bubbles on the surface of the liquid after it has been poured from one vessel into another. To detect it: Boil ten or fifteen cubic centimetres of clear filtered urine in a test-tube and add a drop or two of nitric acid. Heat throws down phosphates, which are dissolved by the nitric acid; the latter throws down urates, but these are dissolved by the heat; when both reagents are applied, nothing remains precipitated but the curdy flakes of albuminous matters.

298. Besides having a high specific gravity, as 1.040 or over, urine which contains *sugar* is passed usually in increased quantities. To ten or fifteen cubic centimetres in a test-tube add a drop or two of a solution of sulphate of copper and an excess of liquor potassæ; then boil the blue solution thus formed, when, if sugar be present, a red suboxid of copper will be precipitated.

299. Urine when freshly passed is clear, but after a time a *sediment* may gather in it. Reddish crystals of *uric acid* may be seen at the bottom of the containing vessel after standing for twenty-four hours. *Urates* are more soluble in warm than in cold water; hence, when present in large quantity, they may become deposited as a fawn-colored powdery sediment after the urine has had time to cool. The urates are in excess when the waste of tissue is great, as in fevers and inflammations, or when the supply of nitrogenous food is greater than is required [240]. *Phosphates* are deposited as the urine becomes alkaline by decomposition, because they are less soluble in an ammoniacal than in an acid urine. They are whitish in color, and may be distinguished from the urates by their failure to dissolve on heating the liquid. They are usually present in excess in cases of nervous depression or deficient vitality. The gradual deposition of these insoluble or sedimentary matters under certain conditions within the bladder gives rise to the formation of *stone* or *gravel*.

300. When a patient does not pass his urine the immediate cause may be either suppression or retention. *Suppression of urine*, in which the kidneys fail to secrete and relieve the blood of urea and other refuse materials, is a grave condition often seen when the secreting tissue is disorganized; the blood becomes poisoned by the retained matters, and, circulating in the brain, gives rise to *uræmic* unconsciousness, convulsions, and death. In *retention of urine* the kidneys continue to purify the blood, but from some local cause the accumulated urine is not expelled from the bladder. It may arise from injury to the spinal cord above the origin of the nerves which supply the bladder, or from some stricture of the urethral passage, or even from an overdistention of the bladder which temporarily paralyzes its muscular coat. Relief in these instances is afforded by the introduction of the catheter. Retention may be present, although the patient's complaint is that of passing water all the time in driblets. The bladder may be so distended that it overflows through the urethra, yet cannot contract for the expulsion of the whole of its contents. This condition is recognized by the tense swelling and feeling of distress in the lower part of the abdomen. In such cases it is better not to empty the bladder at once, but to remove enough to give perceptible relief, and finish the evacuation half an hour afterward.

CHAPTER II.

SHOCK, REACTION, AND INFLAMMATION.

301. When the human body is injured without immediate loss of life, certain effects may usually be observed; and these are: 1, *shock* or *collapse;* 2, *reaction;* and 3, *processes of repair.*

302. SHOCK is a condition of nervous depression following severe injuries. A similar condition caused by mental impressions is generally called *syncope* or *fainting.* The heart beats slowly or quickly, but with so little power that it is unable to force the blood to the brain, as is manifested by sudden pallor of the face, with accompanying giddiness, faintness, noises in the ear, indistinctness of vision, and unconsciousness. The patient should be laid on the ground with his head low; recovery may be hastened by sprinkling a little cold water on the face, offering diluted ammonia or its carbonate for inhalation, and giving a few swallows of cold water or some convenient stimulant.

303. But when shock is the result of injury, it is not so readily controlled. When slight it is characterized by anxiety, tremors, pallor, and faintness, and is relieved by rest in bed, reassuring words, and some mild stimulant, as beef tea, coffee, weak toddy, or aromatic spirit of ammonia in water. When severe, the patient is semi-conscious and incoherent; countenance pale, anxious, and shrunken; surface cold and bedewed with moisture; pulse weak, and respiration irregular and sighing. Recovery is promoted by

placing the patient in bed with his head low, and applying warmth to the extremities and pit of the stomach by hot bottles and friction with warm flannels, while stimulants are given in small quantities at short intervals. When there is reason to fear the existence of internal hemorrhage, as in injuries of the head, chest, and abdomen, alcoholic stimulants should be given with caution.

304. Severe shock may end in insensibility and death by syncope, or in a condition of abnormal excitement. As the duration of this *reaction*, as it is called, is proportioned to the previous depression, it is, of course, desirable to lessen the intensity and duration of the collapse; but this must be effected with caution lest the stimulants used should intensify and prolong the subsequent reaction. This condition is essentially one of fever, and is treated by the means which tend to subdue fever [323]; but as its cause is not an abiding one, it usually subsides within twenty-four hours or becomes merged in that fever which is an accompaniment of the changes that are meanwhile taking place in the injured parts. When, however, the system at the time of the injury is debilitated, low delirium and muscular tremors constitute the characteristic symptoms of a condition which has been named *prostration with excitement*, and which requires for its treatment careful stimulation, with opiates to allay the nervous irritation.

305. REACTION AND INFLAMMATION.—As soon as reaction begins, a local process of repair is set up, which is neither more nor less than an exaggeration of the ordinary nutritive forces of the part. The vessels enlarge and become engorged with blood, making the part assume the local characteristics of inflammation—redness, swelling, heat, and pain. It is red from the accumulated blood; swollen from the engorgement of the vessels and the presence of

the transuded liquid; hot from the destructive oxidation which the flow of oxygenated blood brings upon the injured tissues; and painful from compression of the nerves of sensation by the tumefaction of the parts which surround them. Should the capillary activity subside promptly, leaving no trace of its presence, the part would be said to have been *congested*, as when the ears or fingers suffer temporarily from cold; but should the symptoms persist and a rapid renewal of the deep-seated layers of the cuticle take place, causing the older layers on the surface to crack and peel off, or *desquamate*, an inflammation would then be said to have existed. No well-defined line separates congestion from inflammation; the one is the beginning of the other, and both are essentially an increased activity of the ordinary processes of nutritive change in the injured parts. During congestion, *coagulable lymph* exudes from the vessels, constituting the material of repair. This cements the surfaces of a closed wound and fills up the vacuities of an open one with its granulations. It has but little vitality, and its power of resisting harmful agencies is correspondingly small. When a wound becomes infected by bacterial organisms, as by the introduction of splinters, fragments of clothing, or other foreign matter, or by touching with unclean fingers, or even by exposure to the air, the exuded lymph becomes transformed into a liquid, *pus*, which is of no use as a reparative material, and which drains away as a fetid discharge or is collected in cavities called *abscesses*.

306. Manifestly, after the removal of the cause the indications for treating a congestion are, first, to lessen the quantity of blood in the engorged vessels; and, second, to promote the dispersion of the exudations that have escaped into the injured tissues.

307. The engorged vessels may be relieved by position,

by the application of cold or heat, or by the direct or indirect abstraction of blood.

308. The congested part should be kept at *rest* in an *elevated position;* those who have a felon on the finger speedily discover that hanging the hand low aggravates the suffering, and that raising it on the breast gives measurable relief [344].

309. *Cold,* applied by ice-bags, by evaporating lotions, or by irrigation, acts by diminishing the size of the supplying vessels. Sometimes a cooling astringent, as liquor plumbi subacetatis, is added to the lotion to aid the cold produced by evaporation. *Pressure* is sometimes useful, but usually it is more likely to aggravate the evil by increasing pain, and thus acting as an irritant.

310. *Heat* is applied by *warm-water dressings,* consisting of lint or cotton thoroughly soaked, and covered with oiled silk or gutta-percha cloth to prevent evaporation. *Poultices* are made of bread-crumbs, oatmeal, linseed-meal, or other farinaceous substances. These should be *boiled* for a few minutes, to coagulate all their albuminous constituents and make the poultice lighter than it otherwise would be. Heat, to be efficient, must act on the parts adjoining the injury, relaxing them and preparing them to absorb blood which, without their influence, would go to swell the accumulation in the affected parts. An inflamed finger, for instance, is relieved when the four other fingers and the hand as a whole aid it in disposing of the blood that would otherwise pulsate into its already-swollen tissues.

311. Blood is abstracted by *leeches.* These are placed in a pill-box or wine-glass, which is inverted over the part to which they are to be applied. If it is desirable that a leech should bite on a particular spot, the part should be covered with a piece of blotting-paper perforated at the spot, which should be touched with a drop of blood or milk and water. Leeches fall off when sprinkled with salt, which also makes them disgorge the blood they have swallowed. Used leeches should be kept apart from the others, as they frequently become sick and affect their companions injuriously. Bleeding from leech-bites may be kept up by warm fomentations; it may be stopped by touching the bites with collodion or with a point of lunar caustic.

312. *Cupping-glasses* are also used for the abstraction of blood, but the irritation caused by them necessitates their application at some distance from the congested part. They are most frequently used for the relief of internal congestions, hence the patient must not be incautiously exposed during their application. The surface must be moistened with warm water, that the skin may slide freely under the edge of the cup and rise within it. The cup is held in the left hand, its mouth inclined downward and close to the part on which it is to be placed. The flame of a spirit-lamp is permitted to play in its interior for a second or two; and as the right hand withdraws the lamp, the left sets the cup on the selected spot. Immediately the skin and subjacent tissues rise into the vacuum in a convexity which momentarily becomes more and more turgid with blood. The practice of putting a few drops of alcohol into the cup should not be followed, as the skin may be injured by the burning liquid; and even with the spirit-lamp flame the patient will suffer if the edge of the cup be unnecessarily heated. The pressure of the atmosphere drives the blood from the surrounding tissues into those covered by the partial vacuum of the cups, and the congested spots thus intentionally produced give more or less relief to the congestion under treatment. This, which constitutes *dry* *cupping*, is sometimes all that is needful, the cups being permitted to act for a given time; but when *wet cupping* is desired, the cup first applied is removed as soon as the others are in position, the scarificator is used, and the cup replaced, after which the other cupping-sites are treated, in succession, in like manner. The cup is detached by pressing with the finger on the skin near its edge. The *spring scarificator* should be held firmly in position on the skin when the trigger is pulled; cuts of one-eighth inch generally yield better results than those that are deeper. The operation is finished by cleaning the surface thoroughly, bathing the lancet wounds with an antispetic liquid, and covering them with plaster so applied as to counteract any tendency to gaping of their lips.

313. Counter-irritants act like cupping-glasses on parts adjacent to the congested area. Turpentine stupes, mustard, or ammonia are frequently used over the throat and upper part of the chest in croup; and a large blister is sometimes beneficial over the affected side in congestion of the lung and on the back of the neck when the brain is implicated.

314. Counter-irritants are called *rubefacients* when they redden the skin; *vesicants* when they raise the cuticle by an underlying effusion.

315. The *turpentine stupe* is simply a flannel cloth wrung out of hot water and sprinkled with turpentine.

316. *Mustard*, made into a soft paste with a little tepid water, is spread on a piece of linen or cotton and applied with a piece of tissue-paper, gauze, or fine cambric interposed to prevent the irritant from adhering to the skin; it is kept in position until the surface is reddened, after which a piece of soft lint is used as a protective. When left too long in contact with the skin, mustard causes painful ulcera-

tions. If mild rubefaction only is required, the paste should contain one or two parts of wheat flour.

317. *Ammonia* is commonly employed in the form of the official liniment; or a wine-glass may be filled with pledgets of lint soaked in diluted ammonia, and inverted for one or two minutes over the selected part.

318. A *fly-blister* is made by spreading cerate of cantharides on adhesive plaster, linen, or paper. The skin should be sponged with vinegar before the blister is applied, and there should be no tissue-paper facing such as is used in the case of mustard. The application is removed as soon as vesication takes place, or at the end of ten or twelve hours, even in the absence of vesication. In the latter case, the use of a bread poultice will usually raise the cuticle. Indeed, in children and persons with delicate skins, the substitution of a poultice for the cerate at the expiration of three or four hours is the better practice. If speedy healing be desired, the effused liquid is drained off by one or two punctures and the part dressed with cotton or an antiseptic ointment; if the blister is to be kept open, a circular portion of the cuticle is cut out, but not removed, and the surface dressed with an irritant ointment, as a petrolatum dilution of the ceratum cantharidis. The immediate removal of the cut portion is attended with much pain; it is better, therefore, to permit it to be thrown off naturally.

319. For making small counter-irritant sores, technically called *issues*, potassa or potassa cum calce is generally used. A circular aperture the size of the portion of skin to be inflamed is cut in a piece of adhesive plaster, which is applied to the part. *Potassa* is rubbed on the unprotected spot for a minute, after which the plaster is removed and the part washed with vinegar. *Potassa cum calce* is made into a paste with alcohol and applied for ten or fifteen min-

utes over the aperture, vinegar being used, as in the other case, to wash away the alkali that may be sticking to the destroyed surface. These escharotics excite so much congestive action in the skin as to cause its superficial layer to be thrown off as a slough, leaving an ulcer which may, like a blistered surface, be healed or variously irritated according to the requirements of the case.

320. *Hot water* is readily available as a counter-irritant. Its heat is best applied by means of the head of a hammer or other smooth, metallic body. The hammer, immersed in water at 120° F., dried and held in contact with the skin for two or three seconds, acts as a rubefacient; and if the contact be prolonged for ten seconds, vesication will usually be produced. *Tincture of iodine*, *croton oil*, and *tartar emetic ointment* are frequently used for their counter-irritant properties.

321. Besides these local applications for lessening congestion and removing effusion, general measures are sometimes available and useful. *Purgatives* are valuable in congestive conditions, particularly of the brain. They stimulate the intestinal canal and the organs connected with it; and the afflux of blood to these organs lessens by just so much the engorgement of the blood-vessels in the congested brain. The large watery evacuations produced by Epsom, Glauber, and Rochelle salts relieve the turgescence of the vascular system by withdrawing part of the liquid constituents of the blood, but without inducing that extreme prostration which, in blood-letting, follows a loss of the red corpuscles. The action of *diaphoretics* on the skin and of *diuretics* on the kidneys operates advantageously when the congestion to be relieved does not affect the skin in the one instance or the kidneys in the other. In both there is a temporary afflux of blood to these organs,

which is relieved respectively by a free perspiration or a profuse secretion of urine. *Nauseants*, as tartar emetic and ipecacuanha, operate on the whole system and incidentally only on the congested part. The vital energy of every organ and tissue of the body is diminished that the energy which threatens the congested organ with danger may be lessened.

322. SYMPTOMATIC FEVER.—When an injury is severe, reaction is usually associated with and aggravated by a febrile condition, which is usually referred to as *traumatic, symptomatic,* or *inflammatory.* The skin is hot and dry, the face flushed, eyes injected, pulse quickened, breathing accelerated, tongue white and coated, appetite impaired, except for cooling liquids, bowels constipated, urine scanty and high colored; and there are pains in the loins and limbs, headache, restlessness, confusion of thought, and sometimes delirium, particularly at night. At the onset the patient may feel chilly, but there is never a decreased temperature; on the contrary, even while chilliness is present, the thermometer may show the bodily heat to be two or more degrees above the normal. The febrile heat results from increased activity of the nutritive processes going on in the injured part. The greater heat developed there raises the average temperature of the blood, as the heat of the water-back of a kitchen range raises that of the contents of the hot-water cylinder connected with it. This excess of heat distributed throughout the body stimulates the oxidizing or destructive function of the blood, so that all the tissues are wasted with unusual rapidity, thus adding to the febrile heat and aggravating the abnormal condition of the patient.

323. The measures which are of value in subduing this febrile condition consist of rest, quiet, low diet, and the

agencies already indicated as useful in quieting the local action [306], such as purgatives, nauseants, and diaphoretics, which depress the vitality of the whole system, and certain remedies which reduce the temperature and pulse without causing much general depression.

324. *Rest* has a powerful influence in quieting the heart's action. The pulse is more rapid when one is standing than when sitting; more rapid when sitting than when at rest in bed; and during exercise its rapidity is proportioned to the exertion [186].

325. *Quiet,* which may be regarded as a variety of rest, implies the withdrawal of all stimulating impressions. The patient should be kept in a cool, well-ventilated room, and protected not only from glare, noise, and other physical influences, but as much as possible from all that would interfere with tranquillity of mind.

326. By *low diet* is not meant a restriction to so many ounces per day, for a fevered patient has usually but little desire for food, but the careful avoidance in the diet of everything of a stimulating nature. Milk, eggs, oysters, tea, toast, beef tea, broths, farinas, and jellies constitute its elements.

327. CHRONIC INFLAMMATION.—Sometimes, and chiefly in people of weak vitality, congestion does not entirely subside after the healing of a wound, the filling up of an abscess cavity, or the repair of an injury. The area remains red and somewhat swollen; but as the pressure is slight, there is but little pain; and as the nutritive changes in the part are not active, there is neither local heat nor inflammatory fever. The fault in the constitution, which is the main cause of the persistence of the congestive action, must be remedied by tonics, stimulants, and diet, with appropriate treatment when called for by a *tubercular, rheumatic,* *syphilitic*, or other specific taint in the blood. Local treatment usually consists of stimulant applications to excite the nutritive forces of the part to that energy of action which will remove the exuded matters from a congested area, or repair the damage that may have been caused by suppuration or ulceration.

328. Active exercise promotes the nutritive changes in a healthy part; the same end, in a chronically inflamed part, is induced by what may be called passive exercise. *Friction* and *massage* or *shampooing* are the methods by which this is attempted. The former is a misnomer, for the manipulation intended is not a quick, superficial rubbing that would speedily abrade the skin, but a firm, equable pressure reaching to the deeper tissues and applied in the direction of the circulation so as to aid the onward movement of the blood. The latter adds to the so-called friction a pommelling, kneading, and slapping of the parts, which should never be continued long enough to excite pain. These methods of stimulation are used to remove the exudations which, in sprains and contusions, interfere with the natural movements of the affected parts. *Pressure* enters into these manipulations as a part of the passive exercise; but it is sometimes applied continuously to promote absorption, as by means of a weight or compress on an enlarged gland or *bubo*, or by strapping with strips of plaster in the case of a swollen testicle. The oleaginous substances which enter into the composition of *liniments* are of value mainly in lessening friction and preventing abrasion during the rubbing which is associated with their use; but the turpentine, ammonia, and other stimulants that are frequently employed have each a special value independent of the method of their application. All the substances that are used as *counter-irritants* in acute inflammation [313–

320] may be utilized as local stimulants in chronic cases; but while, as counter-irritants, the site of their application is at some distance from the inflamed part, as stimulants their site is the part itself, or so close to it as to involve it in the stimulant action. Solutions of nitrate of silver, sulphate of zinc, sulphate of copper, and other irritant substances, so diluted as not to over-excite the parts, are in constant use to change the character of the action in indolent sores and in chronic inflammation of the mucous membranes of the eye, throat, urethra, rectum, etc. Tincture of iodine is frequently employed for the dissipation of the effusions and exudations that impede motion after the inflammation which caused them has subsided, because, in addition to its local stimulus, it is taken up into the system, where it acts as a powerful excitant to the processes of absorption; and with the same view iodid of potassium is often administered internally, particularly in scrofulous and rheumatic swellings. In the chronic inflammations and exudations of syphilitic disease mercury is frequently prescribed in small doses to affect the system gradually and gently. Gray powder, the chlorid, and iodids are used internally for this purpose, with or without the external application of mercurial ointment or vapors. A *mercurial vapor bath* is administered by adjusting a blanket around the patient from his neck to the floor as he sits undressed upon a cane-buttoned chair under which, in a porcelain or other suitable vessel, are twenty or thirty grains of calomel. The heat of a spirit-lamp is made to volatilize the calomel, forming a mercurial vapor which condenses on and is absorbed by the skin. The bath lasts fifteen or twenty minutes, and at its conclusion the patient is put to bed wrapped in the blanket which was used to confine the vapors.

CHAPTER III.

SPECIAL INFLAMMATIONS.

329. Burns.—When heat is applied to the body by radiation, as from a glowing fire, some of the minute scales that form the cuticle have their vitality impaired. Congestion follows and the nutritive energies of the part become intensified; but in the course of a few days the action subsides with the completion of a new layer of cuticle and the scaling off or desquamation of the old one. This is known as *rubefaction*—the action produced on the skin by mustard, liniment of ammonia, tincture of iodine, etc., when used as counter-irritants. When produced thus intentionally rubefaction requires no treatment except protection by some soft material; but when it results from accident the congestive reaction may go beyond rubefaction; and as it is undesirable that it should do so, means should be adopted to restrain it. Lead lotion is commonly used for this purpose, with tincture of opium to allay the burning. Exposure to the air increases the congestion and aggravates the pain; hence the value of dusting the part with flour or starch, covering it with cotton-wool, or smearing it with bland unguents, such as palm-oil, oxid of zinc ointment, or petrolatum, to which the addition of carbolic acid is of benefit, as it tends, like opium, to benumb the sensibility; alkaline liquids, as carron oil and strong solutions of carbonate of soda, are also used for this purpose.

330. When the heat is greater, and generally in the injuries called *scalds*, the vitality of the surface layer of the skin is impaired as in rubefaction, bu the reactionary congestion is so much more intense that the effused serum accumulates beneath the injured cuticle, raising it into a blister. This is *vesication*—the action of cantharides, ammonia, acetic acid, etc., when applied for purposes of counter-irritation. Every care must therefore be taken to preserve the injured surface from influences that would intensify the congestion. If the burned surfaces are numerous or extensive they should be uncovered, one by one, or part by part, to lessen exposure and prevent irritation. Blisters, when flaccid, should not be opened, but when distended they should be punctured to relieve tension and lessen the risk of accidental rupture. They are then protected by any of the applications mentioned as useful in the case of rubefaction [329] except flour and starch, which would become caked into an irritant by the exuding liquids; antiseptic absorbent cotton is probably the best; but whatever the application, it should be left undisturbed for several days unless its removal is called for by excessive discharge or putrescent odors. By the end of a week or ten days a delicate new cuticle is formed, over which the old shrivels up and peels off; but the entrance of putrefactive germs from an inefficient protective or antiseptic dressing will lead to *suppuration* if the degenerated matters drain from the surface, to *ulceration* if they are absorbed, and to *sloughing* if they are thrown off as a solid mass. In any of these events the healing of the primary injury may be delayed for weeks. If the whole thickness of the skin is involved in a burn, protective or cooling applications should be used at first to moderate the reactionary congestion; next, soothing poultices or warm-water dressings to promote the ulceration which is to separate the slough; and lastly, the applications called for by the characters of the granulating sore. The cicatrix in burns of this depth becomes powerfully contractile, so that it will in progress of time, and according to its situation, evert the eyelids, distort the features, draw the head to one side or the other, and flex the joints, producing deformities and disabilities that are as difficult to prevent as to cure.

331. In all burns except those of a trivial nature there is more or less disturbance of the general system. Shock [302] is proportioned to the *extent* of the burn as well as to its *intensity*; thus, there may be as complete a collapse in a case of scalding as in a case of charring, if the former involve a larger extent of the surface. Sometimes shock accompanies very trifling burns, in which case it is usually due rather to the circumstances attending the injury than to the injury itself, and is easily removed by stimulants; but when due to the latter it is often prolonged, two or three days elapsing in severe cases before reaction is established.

332. BURNS BY CORROSIVE ACIDS.—The skin when corroded by oil of vitriol, nitric or hydrochloric acid should immediately be washed with a strong solution of carbonate of soda or other alkali, and subsequently treated as a burn. When the eyes are injured, the alkaline solution should contain about ten grains of the carbonate to an ounce of water. Olive or castor oil should afterward be used to protect the disorganized conjunctiva. Stimulants and opium are called for if there is much shock.

333. SCALDS, ETC., OF THE THROAT.—Children are sometimes scalded or burned in the mouth and throat by steam or boiling water, acid, or strongly alkaline liquids, as solutions of ammonia or washing soda. Usually the mucous membrane of the lips and mouth only is injured; but occasionally the larynx is involved. Redness and swelling of the mouth and throat, with difficulty in

swallowing, hoarseness, and accompanying fever, follow the accident. The immediate danger in such cases arises from closure of the rima glottidis [224]. If the local symptoms are due to acids or alkalies, the appropriate antidote [500] should be used as a mouth-wash, and taken into the stomach if there is reason to suppose that any of the dangerous liquid has been swallowed.

334. FROST-BITE.—The frost-bitten part is cold, shrunken, bloodless, and without sensation. The immediate object of treatment is to restore the circulation in the affected region; but the reaction must be cautiously induced lest the subsequent congestion run into gangrene or sloughing. The frozen part should be thawed out by friction with a *mixture of snow or ice and water*. Only when reaction is fairly established should the patient be moved into a warm room.

335. After reaction is established, the lighter grades of frost-bite require no treatment except protection. In cases of greater severity, increasing congestion should be restrained by cooling lotions; but if sloughing threatens, these should be replaced by soothing poultices, with stimulants of resin and turpentine when the action becomes indolent, and charcoal or other deodorizing antiseptics when called for by the gangrenous condition. When the toes, fingers, or larger portions of the extremities are destroyed, they do not become separated from the living parts in such a manner as to leave the latter in good condition for the formation of a rounded stump; on the contrary, as the interior parts suffered less than the exterior from the killing influence of the cold, more of them are preserved; and when the separation takes place they, including the bones, project beyond the level of the superficial tissues that remain. This necessitates surgical interference.

336. CHILBLAINS are chronic congestions produced by cold in constitutions of impaired vitality. The patient should be well fed and exercised to give vigor to the system. This, with the protection of the congested parts from changes of temperature, will alone effect a cure; but local applications, as of camphor, turpentine, or tincture of iodine, are called for, chiefly by the intolerable itching. When sloughing takes place, poultices with resin cerate or turpentine should be used until the surface is ready to granulate.

337. A CONTUSION or bruise is the injury that results from a sudden crushing of the parts. There is swelling, with numbness, dull aching pain, and shock. The swelling results from exudation and extravasation [173], the dark color of the bruise being due to the latter. Where the tissues are lax, as about the eyelids and scrotum, the swelling is greater, in proportion to the injury, than where they are less distensible. Evaporating lotions or ice should be applied to restrain the outflow from the vessels. Ordinarily, in a few days the swelling decreases and the discoloration fades. Absorption is promoted by rubbing equal parts of tincture of arnica and water, with equable pressure, over the contused parts; or soap liniment with tincture of opium may be used if there be much pain. Contusion of the testicle is accompanied with much shock and tendency to syncope. Contusion of the abdomen, as from blows, the kick of a horse, or passage of a wheel over the body, is sometimes followed by nausea, vomiting, and great prostration, although there may be no laceration of the abdominal viscera. In cases of suspected internal injury stimulants should be given with caution, lest hemorrhage be increased. A full opiate dose, the careful removal of the patient to hospital and perfect rest afterward, constitute the treatment of the emergency; purgatives are contraindicated.

338. A SPRAIN is a stretching or tearing of the fibrous bands and expansions that surround a joint or of the muscular and tendinous tissues concerned in the movements of the body. When a sprained joint is examined immediately after the injury, the absence of dislocation or fracture is readily discovered; but if some hours have elapsed, there may be so much pain, swelling, and interference with motion as to render the diagnosis uncertain until the inflammatory action has been to some extent controlled. Evaporating, lead, opiate, arnica, and other lotions have a value only when acting in concert with immobility. Splints and slings should be applied when needful to restrain the joints of the upper extremity; but when the ankle or knee is injured, the patient should be put to bed with the limb elevated on pillows; and after the subsidence of the inflammation, the utmost caution must be exercised in resuming the use of the injured limb. Sprains of the ankle often disable for months on account of chronic inflammation kept up in the joint by its unadvised use. A sprain seems a minor injury, and patients fail to recognize the necessity for absolute rest until they learn by experience that it would perhaps have been better for them if there *had* been a dislocation or fracture, as they would then have submitted to the necessary confinement. Slight sprains of the ankle or knee, associated with little inflammatory action, are sometimes put up in an immovable dressing, as the starch, silicate, or gypsum bandage [440-443], and the patient permitted to move about on crutches; but when the injury is at all severe, the acute inflammation should be controlled before having recourse to these dressings. Stiffness of the joint during convalescence from sprain of the ankle is removed by friction or massage, and effusion by tincture of iodine or other counter-irritants [328].

339. Sprains of the loins are sometimes so severe as to simulate more dangerous injuries, as fracture of the spine; but the patient is usually able, with more or less difficulty, to straighten himself up, and the spinous processes present no irregularity. Blood in these cases sometimes appears in the urine, indicating an injury to the kidneys; but this is not often followed by any inflammation of that organ. Hemorrhage may also take place into the spinal canal, occasioning paralysis of the lower extremities and increasing the resemblance to fracture. Paralysis may also supervene from effusion of serum into the canal, due to the spread of the inflammation from the sprained tissues to the membranes of the cord. The necessity for perfect rest in severe sprains of the loins is therefore imperative. This is aided by suitable measures to allay inflammation [323] in the early period, and counter-irritants and resolvents afterward.

340. ABSCESS.—When the degenerated matter of an exudation accumulates in a liquid form in the interior of a tissue it constitutes an abscess. The walls of the cavity which contains the matter are thickened from exudation, and external to this there is usually a swelling from effusion of serum, and, if the skin be implicated, a redness from congestion. Superficial abscesses are seen in connection with minute punctured wounds from splinters, the presence of which gives rise to the suppuration which eventually casts them out. During the hard stage of the inflammatory swelling, the progress of suppuration should be promoted by large poultices, but as soon as a fluctuating spot is detected, that is, a soft spot which feels to pressure with the finger as if it contained liquid, an incision should be made to liberate the matter. If no incision is made, the abscess cavity becomes enlarged by absorption until the surface is reached;

but in this case the patient has to suffer for a longer period, and owing to the larger size of the cavity more time is required for its healing, and the likelihood of scarring is increased. Granulation is conducted under stimulant, antiseptic, or such other dressings as may be called for by its condition.

341. Boils.—A boil is a hard, knobby, painful swelling developed under the skin, which is more or less congested. It enlarges, and at the end of a week its centre becomes pointed and ruptures, giving issue to a liquid oozing, and showing through the aperture a yellowish slough which is too soft to be pulled out by a forceps, even if it were free from all adhesions to the interior. In a few days the slough breaks down and is discharged, the cavity granulates, and in progress of time the coagulable lymph which formed its walls becomes absorbed. Boils usually come in crops. The treatment for abscess is not suitable here, because the matter cannot be cast out by incision, and poultices increase the size of the interior slough. The boil is best treated by merely protecting it from injury by adhesive plaster, and dressing with resin ointment when the slough is separating.

342. Some large and flat boils are called blind because they do not open to discharge a slough. After increasing for a week or ten days they slowly subside, leaving a hardness at their site for a long time; or, at the period of their culmination, a small purulent point may be discovered, which, instead of opening for the discharge of a slough, dries into a yellowish scale. Poultices and incision are of no benefit. The part should be simply protected by plaster. Any discoverable errors in diet or mode of life should be corrected.

343. Carbuncles are large and tedious boils which occur

generally on the back of the neck in debilitated constitutions. The subcutaneous tissues are involved in the congestion, which is of a livid color and attended with much pain and constitutional irritation. The local treatment consists of poultices or warm-water dressing, with afterward a free crucial incision and applications of resin or turpentine to promote the evacuation of the cavity. Sometimes the opening is made by potassa [319]. If left to itself, the carbuncle is opened by ulceration at two or more places, which uncover the underlying slough as they increase in size. A nutritious diet is always, and stimulants generally, required to effect the cure.

344. In a WHITLOW or FELON, which is usually occasioned by a prick on the end of a finger, the exudation is bound down by the strong membrane covering the bone or the fibrous tissues which hold the tendons of the finger in their place. The consequent pressure is a source of severe throbbing pain even when there is but little redness or swelling. The patient is deprived of sleep, and his whole system disturbed, by the intensity of his suffering. The hand should be enveloped in a large poultice, and kept in an elevated position; opium or Dover's powder should be given to allay pain and promote the subsidence of the congestion; but if the symptoms are not favorably impressed within twenty-four hours, a free cut should be made *down to the bone* to relieve alike the tension in the part and the suffering of the patient. To wait until the accumulated matter reached the surface by ulcerative action would be to invite the destruction of the part and many weeks of disability, as the matter finds it easier to spread upward into the palm of the hand along the tendons than to come to the surface.

345. A GUMBOIL is a suppuration connected with the fang of a tooth. A congestion with a slight exudation of lymph occurs at the deep end of the fang. As this lymph is locked in on all sides by bone, there is intense pain, which soon becomes throbbing. If at this stage the tooth be extracted, the matter escapes and the trouble is at an end; but it is not always advisable to sacrifice a tooth on account of the temporary inflammation at its root. Occasionally the matter finds its way between the fang and the socket, and escapes gradually at the neck of the tooth. The usual course, however, is for the matter, by its pressure, to cause absorption of the bony tissue in which it is confined, until an aperture is made which permits it to escape and accumulate between the surface of the jaw-bone and the gum. Accordingly, when any soft, boggy swelling is observed on the gum, it should be immediately incised to permit the matter to escape. For the time being the gumboil is at an end; the swelling subsides and the incision heals; but very frequently the healing is not perfect. There often remains a minute aperture on the site of the incision, and from it matter occasionally exudes. Sometimes the aperture becomes blocked up, and there may be a little pain and swelling before the accumulated matter finds its way through the fistulous opening. To effect a cure in this instance, the decayed root must be removed. If the matter of the original gumboil be not liberated when it has penetrated through the bone and reached the gum, it will generally cause much swelling, and finally break within the mouth by the thinning of the walls of the abscess cavity; but sometimes the matter dissects its way into the tissues of the cheek, causing much swelling of the face and subsequent disfigurement by the scar of its opening. This should be avoided by the liberation of the matter while it is yet confined between the gum and the bone.

346. Ulceration of a bone is called *caries*. The same

term is applied to the softening and absorption by which cavities are formed in the dentine [208] of the teeth. *Toothache* is one of the symptoms of this decay. When the cavity is small and does not penetrate to the dental pulp, it should be carefully cleaned out with absorbent cotton on the end of a suitable instrument, and then plugged with cotton moistened with carbolic acid, to allay sensitiveness. Afterwards, when the cavity has lost its tenderness, it should be dried and filled with gutta-percha, softened by a heat which should not be greater than the skin of the wrist or back of the hand can bear without pain. When the cavity is large and the pulp exposed, as evidenced by the extreme sensitiveness of the bottom of the cavity, the relief of the toothache requires the *extraction of the tooth* or the *killing of the nerve*. To effect the latter purpose, a mixture of equal parts of laundry soap and arsenious acid is used. The cavity having been carefully dried, a small piece, about the size of a pin-head, of the arsenical soap is introduced into it and held in position for twenty-four hours by a filling of cotton-wool. This intensifies the pain for several hours, but afterward relief is experienced. Care should be taken that the action of the arsenic is confined to the cavity. Carbolic acid on cotton-wool is used to allay any remaining tenderness, after which the cavity may be plugged with gutta-percha, as a preservative, until the services of a dentist are obtained.

347. When a tooth has to be sacrificed because of a gumboil or a large cavity of decay, a suitable tooth-forceps should be selected. *Forceps* are made of different sizes and shapes; but every one should be strong and unyielding in its jaws and handles, and should move easily at its joint, without being in the slightest degree loose-jointed; the jaws should fit accurately to the neck of the tooth to be extracted, and should flare out toward their joint, so as to cap the crown of the tooth without compressing it. The jaws of an upper-tooth forceps are usually in line with the handles, while those of the lower-tooth forceps are set at a considerable angle. Forceps for single-fanged teeth [207] have the extremity of their jaws gently curved and plain; those for the upper molars have a ridge on the face of the outer blade for catching hold of the crotch between the outer fangs; those for the lower molars have a similar ridge on both blades to fit on each side into the crotch; a pointed *cow-horn* forceps is designed for lifting out the lower molars. The various forceps at the command of the intending operator should be carefully studied as regards their adjustment to the teeth and the direction of the traction to be employed. The operating-chair should be solid, with a convenient grasp for the hands of the patient, and a low neck-rest, that his head may be thrown well back.

348. The operator stands on the *right side* when an upper tooth is to be extracted; on the *left* when a left lower tooth is the subject; and *behind*, operating over the face of the patient, when a right lower tooth is to be removed. The fingers of his left hand should steady the jaw near the affected tooth and afford counter-force during the extraction. In laying hold, the blades are closed lightly on the tooth and the points are pushed well home on the neck beneath the gum, keeping their long axis in line with that of the tooth. The blades are then closed firmly on the tooth so as not to slip, yet not to crush in the sides of a hollowed tooth. A steadily increasing pull is then made, with a slight rotary movement if the tooth is single-fanged, and an equally slight lateral or forward and back movement, when the tooth will come away. In extracting molars, the traction is combined only with an outward and inward movement; any attempt at rotation would break the fangs. Care should be taken that the extracted tooth do not slip from the forceps into the throat of the patient.

349. If the tooth break during the attempt at extraction an effort may be made to remove the stump, but it should not be prolonged if not promptly successful. The patient may be solaced with the information that his toothache will subside, as the dental pulp has now been removed, or, if the relief of a gumboil was intended, that the matter will now find exit by the central canal and sides of the fang. Stumps rarely call for extraction save at the hands of a dentist to prepare the gums for artificial teeth. As the dental nerve is destroyed, they are seldom the subject of toothache. A steel lever is useful in their extraction, as is also a shorter gouge-like instrument; these, which are called *elevators*, are inserted between the socket and the stump; but the greatest care must be taken lest they slip and injure the mouth, tongue, or throat.

350. A small dental case is provided by the Medical Department for use at military posts where the services of a qualified dentist cannot be obtained. By its means a careful and neat-handed operator can insert a stopping that will last for years. It contains a *mirror* for examinations, a *scaler* for removing tartar, an *explorer* for examining cavities, *chisels* for trimming their edges, and *excavators* for cutting out decayed dentine, with forceps for handling absorbents, etc., and a supply of prepared gutta-percha. Heavy socket-handles are provided for holding the instruments. A *spatula* is used in filling large cavities, and *burnishers* when the cavities are small. The gutta-percha stopping requires to be cut into pellets proportioned to the size of the cavity to be filled. These are warmed on a piece of porcelain over a lamp-flame, taking care not to overheat them. They are then worked into the cleaned and dried cavity with warm instruments. Moisture in the cavity will defeat the intention of the operation.

351. CORNS.—Continued pressure causes absorption and ulceration; intermittent pressure stimulates growth. Corns are an illustration of the effects of intermittent pressure; but it often happens that the stimulation induces inflammation and suppuration. When acutely inflamed, the part should be poulticed or treated with water dressing, covered with oiled silk, and subsequently punctured to liberate the matter, removing the corn at the same time, if this can be readily accomplished. The proper treatment for uninflamed corns, whether *hard*, as over the prominences of the toes, or *soft*, as on their contiguous surfaces, is to keep them closely trimmed, particularly toward the centre, and to wear shoes that do not chafe or press upon the part or crowd the toes together. Care of the feet is of greater benefit in the cure of corns than more energetic and painful methods of eradication.

352. BUNIONS are caused by the pressure of narrow-toed shoes. Naturally, the toes spread away from each other, but the narrow-toed shoe crowds them on the middle toe as a centre, the great toe being deflected outward and the little toe inward, while the root of each forms a prominent angle on the side of the foot. This angle, unsupported by adjacent structures, receives the pressure of the boot or shoe, and its skin becomes first thickened by the stimulation, and then painfully congested by its continuance. If this protest against the unsuitable footgear be disregarded, effusion takes place beneath the skin and forms a water-pad, as it were, for the protection of the joint. Under a continuance of the pressure the effused serum becomes replaced by purulent matter, which escapes by the ulceration of a

small aperture through the overlying cuticle. The cure of the bunion requires the removal of the cause, the relief of the congestion by a poultice, the absorption of the effusion by iodine paint, or the granulation of the suppurating cavity under appropriate dressings—soothing, if the granulations be unprogressive from inflammation, and stimulating, if unprogressive from indolence.

353. BLISTERS ON THE FEET.—In the early days of a march soldiers sometimes suffer from blistering of the feet, particularly the heel, developed by chafing in ill-fitting shoes or such as have not been broken in to the form of the feet. The blister should be drained and protected next day by a piece of rubber plaster.

354. CHAFING IN THE GROINS is also a frequent annoyance on the march. It should be treated in camp by washing the excoriated surface and applying a lotion of acetate of lead, and on the march by the use of oxid of zinc ointment.

CHAPTER IV.

WOUNDS.

355. A WOUND is an injury inflicted by mechanical violence on the living body—hence a bruise is a wound; but the term is usually restricted to a separation of parts. An *incised* wound is one made by a cutting instrument; in a *lacerated* wound the tissues are torn apart; and in a *contused* wound they are broken by some irregularity on the surface of the bruising instrument or by the violence of its application. Sometimes a contused wound presents the appearance of an incision, as when the scalp is split open by a club. A *punctured* wound has great depth as compared with its other measurements. A *bayonet wound* is a characteristic punctured wound; a *sabre cut* is a contused incision; *gunshot wounds* are contused lacerations.

INCISED WOUNDS.

356. The requirements of treatment in the case of wounds are: To arrest hemorrhage; to remove shock; and to adjust the separated surfaces and to so retain them until the exuded lymph or granulations have acquired strength enough to withstand the separating tendencies of the ordinary movements of the part. Close contact cannot be effected if any foreign matter, such as splinters of the weapon, sand, or earthy particles, fragments of clothing, or even coagulated blood be lodged in the wound. These must be picked away by the forceps or fingers, or washed

off with a plentiful supply of water. The skin of the neighboring surface should also be cleaned, and, if necessary, shaved. The part is then placed in that position which will relax the injured tissues and permit the sides of the wound to be brought together, and the edges are stitched at half-inch or other suitable intervals if plaster alone will not hold them in position. A slightly curved needle is pushed through the skin at about one-fourth of an inch from the edge of the wound, and thence through the opposite tissues,

Closing a wound by suture.

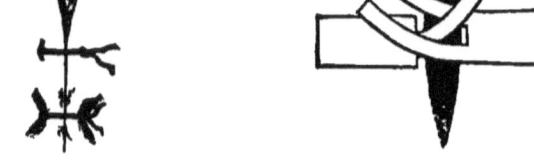

Securing apposition by plaster strips.

drawing after it the suture thread, which is then tied in a reef or surgeon's knot [389] and cut off short.

357. Strips of plaster are applied across the line of the wound at short intervals or between the sutures, if any have been inserted. They should be long enough to extend several inches on each side of the wound, and they hold better when slashed two or three times with the scissors for half an inch longitudinally at each end. In applying each, and particularly the first one, which should be across the centre of the line of wound, one-half of the strip should be placed in position on one side of the wound; and when it adheres firmly, traction should be made on the unapplied half with one hand while the other draws toward the wound the tissues on which the plaster is to be laid, that it, when applied, may the better retain the sides of the wound in apposition. A very good way of applying plaster strips is shown in the accompanying illustration. When a clean incised wound has been drawn together by plaster or sutures, the only dressing required is a protection from the air and from changes of temperature. Some folds of lint or sublimated gauze may be laid over it and retained by a bandage; and if the wound is situated near a joint, the occasional movement of which would tend to draw the edges apart, a suitable splint should be applied to restrain its motion. A *cradle* or light curved framework is sometimes placed over a wound to relieve it from the weight of the blankets. The sides of such a wound will in forty-eight hours be so firmly united that each suture may be snipped at the side of its knot and removed by seizing the knot between the points of a dressing forceps; but the plaster and splint should be left until the union is strong enough to sustain the ordinary strain incidental to their situation. In removing the strips of plaster, each should be raised from one end to the line of the wound, then from the other end to the same line, after which its middle may be detached; in a large wound the centre strip should be replaced by a fresh one before the support of the others is withdrawn. Wounds which progress thus favorably are said to heal by *primary union* or by the *first intention.*

Lacerated Wounds.

358. It is sometimes difficult to bring the sides of a lacerated wound into that close contact which is needful to primary union. When the edges are brought together, there may be interstices and cavities in the interior which the most carefully arranged compresses will fail to obliterate. Moreover, there may be loss of substance, not only in

the interior, but of the surface, which will prevent the edges from coming together. Such a wound does not bleed so much as an incised one, but it requires the exercise of greater care to secure the freedom of its irregular surfaces from the presence of foreign matter. After it has been thoroughly cleaned the wound should be closed by stitches and slips of plaster, with small pads or *compresses* of folded lint to preserve its sides in contact, and a splint, if needful, to restrain motion. The object is to obtain primary union of as much of the wound as possible.

359. The presence of lymph or of a thin layer of coagulated blood between the surfaces does not interfere with this union; but when there are considerable cavities within the wound, union in this adhesive way does not take place. The lymph lying next to the living surface becomes converted into living *granulations,* and that which is more remote, lying in the centre of the cavity or interstice, breaks down into a liquid which drains from the wound. The formation of granulations progresses gradually, until ultimately the two sides come into apposition, and adhesion takes place by what is called *secondary union* in contradistinction to that by primary adhesion.

360. When the loss of tissue involves the superficial parts, constituting what is called an open wound, repair is effected by granulation; and when this process has brought the reparative material in the wound up to the level of the surrounding surface, a delicate skin, soon becoming thicker and stronger, spreads from the edges over the surface and completes what is called the *cicatrization* of the wound.

361. If, however, the granulations be injudiciously treated by warm, moist, and relaxant applications, such as poultices, they may lose their fresh red color and their power of exuding adhesive material; so that, when opposite surfaces come together in the interior, they are unable to unite, and when the proper level in the open wound is reached, they fail to cicatrize, rising, instead, above the level as a fungous growth, constituting what has been called *proud flesh*. The sluggish granulations must be touched freely with lunar caustic or sulphate of copper. This destroys their superficial parts, and the action set up in the capillaries of the deeper parts in order to throw off the cauterized slough leaves them in fitter condition for adhesion or cicatrization, as the case may be.

362. When a wound is infected, owing to imperfect cleansing at first, to subsequent injudicious handling, or to exposure, it becomes red, hot, swollen, and painful, and the discharge from it profuse and acrid—in fact, *it festers;* and if the swelling should block up the track by which the discharges escape, or if the wound itself be of such a character that the generated matter tends to become bagged in some cavity or infiltrated in the subcutaneous or intermuscular cellular tissue, a dangerous inflammation will result. The wound should be freed from all putrefying matters; and if this cannot be accomplished by flushing it with disinfecting solutions, a special opening should be made to facilitate their escape. When it has been thoroughly cleaned it may be again dressed with antiseptics, which are to be renewed as soon as materially contaminated with the discharges. *Irrigation*, which was formerly employed as a method of continuously cooling an inflamed wound, is one of the means by which the putrefactive tendency may be neutralized. It is effected by placing a vessel containing a supply of the liquid to be used on a plane a foot or two higher than the wound and siphoning its contents through a rubber tube, delivering in or on the wound. The flow through the tube

must be regulated by a clip or some extemporized means of pressure.

Contused Wounds.

363. A part which has been severely contused may be several days or weeks before the discoloration consequent on extravasation of blood is effaced and the tissues return to their healthy condition [337]. So in a contused wound time must elapse before the severed surfaces are in a condition to unite. There can therefore be no union by first intention, and sutures are useless in view of the swelling that would necessitate their removal in the course of a few hours as reaction progresses. Plaster strips suffice to preserve adjustment after the wound has been cleaned. On account of the lowered vitality of the injured tissues it is of importance to thoroughly disinfect and protect a severe contused wound. Without an efficient antiseptic treatment, the germs of putrefaction would find in the extravasations and exudations of the injured tissues the most favorable conditions for their growth and propagation, and, as a result, a violent action would be set up, attended with much loss of substance by sloughing, a rapid spread of the inflammation by the acrid products, and the danger of constitutional infection by their absorption into the blood. On the other hand, with an efficient antisepsis the injured tissues break down and are absorbed or drain away, leaving the wound free to granulate and heal like the cavities and interstices of a wound that is merely lacerated.

Gunshot Wounds.

364. Gunshot wounds are treated like other contused lacerations; but the battlefield produces them in such numbers as to render imperative the adoption of a system for their satisfactory management.

365. First Aid concerns itself with shock, hemorrhage, and the application of protective dressings.

366. Shock is, as in other wounds, usually proportioned to the severity of the injury; but sometimes the excitement of victorious battle counteracts the depressing influences of a serious wound, while, when the mind is affected by defeat and the body exhausted by fatigue, a comparatively trivial injury may exercise a powerful impression. When shock is severe, place the patient on his back with his head low; check hemorrhage, if it exists; loosen belts and clothing, and give 5 c.c. of aromatic spirit of ammonia or 20 c.c. of whiskey, with cheering words before removal from the field.

367. Bleeding. — Oozing is usually restrained by the dressing which is intended for the protection of the wound. Venous bleeding is suppressed by graduated compresses [384] on the wound; and if from a limb, by a bandage applied firmly from the toes or fingers up to and over the compresses. Arterial bleeding is controlled by compresses, in wounds of the head, hand, and foot; but when the jet comes from the thigh, leg, arm, or forearm, a tourniquet or rubber bandage should be applied to the main arterial trunk [387]. If the blood springs from the neck or the walls of the chest or abdomen, torsion [386] or the ligature [388] is used; failing this, efficient pressure with the fingers must be employed until the hospital is reached. The *common carotid* is reached by pressing the finger deeply into the neck in a backward and inward direction at the anterior margin of the sternomastoid muscle; the *facial* as it curves from the neck to the face over the base of the lower jaw about an inch in front of its angle; the *subclavian* may be flattened against the first rib by pressing the thumb downward into the hollow behind the collar-bone; the *axillary* by placing the thumb in the armpit and pressing outward against the humerus; the *brachial* in the upper part of its course by a similar outward pressure, but immediately above the elbow the pressure must be from before backward; the *radial* and *ulnar* arteries are reached by the fingers along the outer and inner sides of the forearm near the wrist; and bleeding from the *palmar* arches is controlled by graduated compresses backed by wooden splints. The *femoral* artery is reached below the middle of the groin by pressure with the thumb backward against the femur; the *popliteal* by pressure from behind against the knee-joint; the *anterior tibial* in the lower part of its course by pressure along a line drawn from the head of the fibula to midway between the malleoli; the *dorsal artery* of the foot along the outer side of the extensor tendon of the great toe; the *posterior tibial* in the hollow between the inner ankle and the heel.

368. When the hemorrhage has been very great, the patient is completely prostrated. His face and lips are pale, countenance shrunken, skin cold and bedewed with drops of perspiration; his pulse is rapid, small, weak, and fluttering, perhaps almost imperceptible; his breathing quiet, but frequently interrupted by a long, sighing expiration; his voice is weak and whispering, and his limbs powerless or at times tossed about in sudden, aimless movements. If he is conscious, he calls for water to allay his thirst; but he is often unconscious or unobservant of what is passing, muttering occasionally to himself in a low delirium. A case of this kind, occurring in the field, must be carried with the utmost care by hand-litter to the hospital, administering stimulants by the way, and continuing the treatment on the litter until the returning strength of the patient authorizes his removal to bed.

369. The object of a PRIMARY or FIELD DRESSING is to protect the wound. The contents of a first-aid packet should be used in accordance with its printed directions, or the wound should be covered with layers of sublimated lint, cotton, jute, etc., and secured with plaster or the triangular bandage; but if the wound is already soiled it should be freed from foreign matters and be made antiseptically clean before being dressed.

370. The proper size for a *triangular bandage* is from forty to fifty inches along the base and twenty to twenty-four inches from the centre of the base to the apex. A piece of linen, muslin, or other suitable material of this shape and size fulfils many of the requirements of *first aid* in surgical emergencies. It may be folded in the form of a cravat for application to the head, neck, or limbs, as well as for fixing splints or other protective appliances. In bandaging the *hand*, the injured member is laid upon the open triangle, the fingers toward the apex, and the wrist or forearm over the middle of the base; the apex is then turned upward over the hand, and held in position by crossing the ends of the bandage over it, and tying them around the wrist. The *foot* is bandaged in a similar manner, the ends being secured around the ankle. On the same principle the *stump after* the *amputation* of a limb may be treated or dressings retained upon any part of the *head*. It may be applied to the *chest* by fastening the ends around the body and drawing the apex over one or the other shoulder down in front or behind to be made fast to the girdle formed by its base; to the *pelvis* by encircling the body and bringing the apex up between the thighs in front or behind to be fastened to the girdle. For the *shoulder* a strip may be torn lengthwise from the base of the triangle, passed around the neck on the injured side, and fastened in the uninjured armpit; the small triangle remaining is made fast by its base around the arm of the injured side, and its apex is carried up over the shoulder to be secured to the strip already applied. For wounds of the *hip* or *thigh* two triangles are required—one is folded and applied as a girdle around the body, the base of the other is passed around the thigh, and its apex brought up and made fast to the girdle. As a sling for an injured upper extremity one-half is laid over the breast and shoulder of the uninjured side, so as to bring the midlength of its base under the hand and its apex beyond the elbow to be supported; the other half is carried up over the shoulder of the injured side and its ends are tied on the nape of the neck. The point or apex of the triangle is then smoothed out, folded snugly around the elbow, and made fast.

Sling for forearm.

371. When bones are fractured, place the limb in as natural and easy a position as possible and apply a bandage or padded splints for comfort during transportation. Rifles, swords, scabbards, pieces of wood, lath, or shingles, branches or twigs, may be used as splints, and clothing, blankets, hay, straw, grass, or moss as padding. *Collar-bone* or *shoulder-blade:* put the forearm and hand in a sling and bind the arm to the body. *Humerus:* apply two splints, one in front and one behind, when the break is in the lower part; otherwise, on the inner and outer sides; and support the arm in a sling. *Forearm:* place thumb up and apply a splint along the outer surface to the wrist and along the inner surface to the ends of the fingers; support by sling. *Bones of leg and thigh:* apply splints on each side. For the thigh the outside splint should extend from the armpit to beyond the foot. A blanket made into two rolls, with a trough for the limb between them, is useful. Bind the injured limb to the sound one.

Infected Wounds.

372. Wounds are occasionally assailed by germs of greater virulence than those of putrefaction, particularly when many patients are crowded together under the same roof. ERYSIPELAS and HOSPITAL GANGRENE both result from the infection of the wound by specific germs through the medium of a foul atmosphere, tainted fingers, instruments, sponges, etc. Each of these germs produces its respective disease: the one a violent inflammation, which spreads far from the original wound along the skin and its underlying tissues, causing the accumulation of purulent matters at various points, with much disorganization of the affected parts and a corresponding amount of febrile action; the other an equally violent inflammation which, although spreading with less rapidity, exercises a more deadly action on the parts involved, causing their mortification or death, and is associated with a low or typhoid form of fever. To avoid these infections, the wounded should be treated in the open air or in tents rather than in crowded rooms.

373. ARROW WOUNDS are punctured wounds with more or less of contusion, but they differ from bullet penetrations in being always infected with germs. Both the arrowhead and its shaft are always unclean; the former is often made fast with animal fibres, and the latter painted with blood or vegetable juices. Wounds penetrating the cranium, chest, or abdomen are therefore exceedingly dangerous. The head of the arrow should be removed, if possible; but the projecting flukes make this difficult. First aid is restricted to cutting away the shaft a few inches outside the entrance and taking care that no movement of the patient shall bury the head deeper in the tissues. But the sooner the effort is made to remove the arrow-head the greater will be the chances of success, for the fastenings of many arrow-heads become loosened in the moist tissues, so that traction on the shank removes it, leaving the head within. If needful, the wound must be enlarged for the passage of a long-limbed forceps, and only when the head has been secured between the blades should traction be made on the shank and forceps as a whole. Cases may occur when the best, if not the only possible, way of removing the missile is to push it onward in its line of direction, meeting it, for extraction, with an incision through the tissues which it must of necessity penetrate. Sometimes, in the excitement of battle with Indian foes, the wounded man will seize the shaft of the arrow and drag it from the wound, in which case the head is usually left in the tissues, and the difficulties attending its removal are much increased.

374. Wounds which become infected with the germs of erysipelas or hospital gangrene [372] are in truth POISONED WOUNDS; but this title is usually reserved for those that are inoculated with the poisonous agent by the instrument which inflicted them. The scratches or punctures accidentally received in examining or dissecting the dead body, or in operating on the living, are often poisoned as well as those caused by the sting of a hornet or nettle, the bite of a rabid dog, or the fangs of a venomous serpent. The gravity of these depends on the nature of the virus or germ introduced. The wound in itself is generally a minor consideration—a mere abrasion often proving more dangerous than a free laceration, as the oozing of blood from the latter tends to the removal of the infectious material.

375. Dissection wounds sometimes heal as readily as if uninfected; in other cases the scratch inflames and suppurates, leaving an ulcer which heals by granulation or by forming a thick scab, as in the case of the vaccine virus; in a third set of cases the inflammatory action is not confined to the wound, but involves all the neighboring tissues, as of the finger and hand, while red lines of inflamed lymphatic vessels reach upward to the armpit, the glands of which become implicated in the swelling and suppuration. Dissection wounds are especially dangerous when the virus implanted is derived from a patient who has died of blood-poisoning, erysipelas, or inflammation of the peritoneum or pleura. The fever which accompanies them is usually of a low or typhoid type. The treatment of dissection, as of all poisoned wounds, looks to the removal or destruction of the poison before it has time to be absorbed. Prompt action is therefore essential. A constriction should be placed around the finger or hand above the injured point, which should be immediately washed in a germicidal liquid, incised to promote bleeding, and sucked vigorously to remove the septic matter. After this, if the matter is known to be of a specially dangerous character, lunar caustic should be freely applied and the part wrapped up in a large soothing poultice. In the absence of nitrate of silver, any powerful escharotic may be employed, as the caustic alkalies, sulphuric or nitric acid applied by means of a wooden probe, a drop of boiling oil, or a small fragment of chlorid of zinc paste made by moistening and mixing one part of the salt with two of flour.

376. The BITE OF A RABID DOG should be treated as a dangerous dissection wound. When lunar caustic is not at hand, the wound, after laceration, free bleeding, and vigorous suction, should be cauterized with a steel rod, probe, or other suitable instrument, at a dull red heat. Experience has shown the efficiency of solid nitrate of silver, when promptly and freely applied; but the terrible and deadly nature of hydrophobia has begotten a distrust of any but the most heroic measures. Hence a thorough excision of the bitten part is by some recommended, to be followed by caustic or the actual cautery.

377. SERPENT BITES must be promptly sucked, excised, and cauterized, an encircling ligature meanwhile having been placed on the cardiac side of the wound; some practitioners have used alcohol, ammonia, or tincture of iodine locally to neutralize or decompose the virus. The effect of the poison depends on its virulence in the individual case. Sometimes the part inflames, the inflammation spreading like erysipelas along the subcutaneous cellular tissue, and affecting the glands by way of the lymphatic vessels. The patient may die after several weeks of suppuration and suffering, or recover after many months with a limb more or less impaired by destructive inflammation. In the most virulent cases, the patient dies in a few hours from disorganization of the blood, the body becoming swollen, ecchymosed, and undergoing rapid decomposition. If there is much depression during and after the local treatment, alcoholic stimulants should be freely administered.

378. The BITES OF TARANTULAS and the STINGS OF SCORPIONS, CENTIPEDES, ETC., are frequently spoken of in the Western service of our army, but are rarely seen. When they occur, they must be treated on the same general principles as other wounds of this class. The application of diluted ammonia or liquor potassæ to the injured point is recommended by some for the neutralization of the poison.

379. The trifling wounds caused by the STINGS OF BEES OR WASPS are best treated by suction, which usually withdraws the sting, or touching with diluted ammonia, after which the local inflammation is subdued by means of lead lotion or cold-water dressings.

380. GLANDERS.—The inflammatory discharges from a glandered horse or mule produce disease in man when they come in contact with a wound, an abraded skin, or even with the unbroken mucous membrane of the nose or lips. The contagious nature and the invariable fatality of the disease call for the utmost care in dealing with its suspected occurrence in the horse. The earliest sign consists of a steadily flowing watery or slightly mucous discharge from the nostrils, with swelling of the glands under the jaw of the affected side. This may continue unchanged for several months, the mucous membrane being dark purplish instead of faintly pink as in health or bright red as in ordinary catarrhal inflammation; but usually the discharge becomes thick, viscid, discolored, bloody, and offensive, the mucous membrane ulcerated, and the lymphatic glands of the face tumefied or transformed into putrid sores. The glands on the inside of the thighs are next involved, and speedily the whole body is converted into a mass of putrefaction. In some cases the disease begins by swelling and ulceration of a few of the glands of the extremities; the lymphatic vessels connected with these become swollen and corded, and implicate the other glands with which they communicate. These knots of enlarged glands, called farcy buds, ulcerate in their turn, and the matter discharged from them is as infectious and dangerous as that from the glandered mucous membrane. In man, the two modifications of the disease are usually combined; whether the disease begins as glanders or as farcy, the other modification is speedily developed.

381. When a suspicious case appears in a stable, the

affected animal should be isolated and cared for until expert testimony has decided upon its destruction as a dangerous focus of infection, or its treatment,—for farcy in its early stages has been successfully treated. Those having care of the animal should not sleep under the same shelter with their charge; all unnecessary contact should be avoided, and after unavoidable contact, either with the animal or surrounding objects which might be infected, the greatest care should be given to washing away every trace of such contact; the danger in the breath of the animal should be fully appreciated, and exposure to its impetus during the sneezing or coughing, which frequently occurs, should especially be shunned. Disinfectant solutions [536] should be freely used.

CHAPTER V.

HEMORRHAGE FROM WOUNDS.

382. CAPILLARY HEMORRHAGE consists of a general oozing of red blood from the surface of the wound. It is controlled by clearing away all clots, freely exposing the part to the air, and elevating it; by affusion with cold or hot water [275] or the application of ice; by the use of styptics, which coagulate the blood in the cut ends of the vessels, and by bringing the sides of the wound together when they can be so bandaged as to prevent the bagging of accumulating blood.

383. Liquor ferri subsulphatis, or Monsel's solution, is the styptic commonly used; the persulphate is too irritant for this purpose. It is applied locally by means of a dossil of lint; it is inhaled from an atomizer in bleeding from the lungs, and administered in 3 to 6 minim doses in hemorrhage from the stomach or bowels. Tannin is applied locally in solutions of suitable strength; dissolved in alcohol and ether and mixed with collodion, it forms a styptic collodion used for sealing the edges of wounds. Alum in solution, turpentine, tincture of catechu, tincture or infusion of matico, and other astringents are also frequently employed by mopping the bleeding sufrace with a pledget of lint or cotton which has been soaked in the styptic liquid. *Epistaxis* or bleeding from the nose is treated by any of these methods aided by compression of the facial artery [367]. Ordinarily in full-blooded individuals nose-bleeding is salutary and easily stopped; but in broken-down constitutions and in heart disease it is sometimes persistent and dangerous. Bleeding from the *socket of an extracted tooth* may be controlled by removing all clots from the cavity, packing into it, bit by bit, a long strip of dry lint, a quarter of an inch wide, placing a small-folded compress over this, and keeping the jaws tightly closed on the whole for several hours.

384. VENOUS HEMORRHAGE is dark-colored, and flows in a steady stream. Pressure with the finger or with graduated compresses is the best means of controlling this species of hemorrhage. Graduated compresses consist of small pads of folded lint, cotton, or other suitable material. The smallest is placed on the bleeding-point; others are laid over it, the outermost being the largest, so that when the whole is bound in position by means of a bandage the pressure is transmitted to the bleeding vessels. If the bleeding is from a limb, a bandage should be applied from the toes or fingers to the level of the wound, prior to placing the compresses in position. A tight bandage around the limb above the wound increases venous bleeding.

A graduated compress.

385. ARTERIAL HEMORRHAGE is florid red in color and escapes from the wound in spurts or jets. A small arterial point may be recognized on a profusely oozing surface by the impulse of the escaping blood and the brightness of its color. Nature operates in three ways to prevent fatal consequences from the division of these vessels: by the contraction of the divided ends, the coagulation of the blood, and the influence of the injury on the heart and nervous system. When an artery is divided its muscular coat contracts on the instant, narrowing and withdrawing

its cut end into the tube of the outer coat, and carrying the interior lining along with it. A gash with a sharp knife is followed by a spirt of blood which escapes before the inner coats have had time to contract; but this immediately subsides to a capillary oozing unless some comparatively large vessel has been wounded. When the cutting instrument is blunt the bleeding is relatively less, because the irritation caused by the coarse edge stimulates the vessels to contract. Some vascular tumors which could not be excised on account of hemorrhage have been successfully removed by the *écraseur*, a strong wire cord which is looped over the tumor and made to cut its way through. Contused and lacerated wounds bleed less than incised wounds: an arm has been torn off by machinery without notable hemorrhage from the lacerated axillary artery. The closing of the divided vessels, which is begun by the contraction of the interior coats, is completed by the formation within them of a plug of coagulated blood; for the retracted and roughened ends favor the deposition of coagulum. The tendency to fainting [302] which follows a wound conduces to the suppression of bleeding by diminishing the force of the heart's action.

386. Artificial means for arresting hemorrhage are based upon those employed by nature. *Douching* with cold or hot water, or *exposure to the air*, stimulates the cut vessels to greater contraction, and is suitable to the cases of small vessels that do little more than ooze. *Continued pressure* with the finger upon a bleeding-point will restrain hemorrhage until the vessel becomes plugged with coagulum; this pressure may be combined with the use of *styptics*. When the spurting vessels are small, bleeding may often be stopped by *torsion*, by which is meant seizing the bleeding-point between the blades of a forceps and twisting it three or four times, thus converting that particular part of the wound into a laceration, which does not bleed. When the vessel is larger and yields a dangerous jet, the finger must be placed upon it to prevent loss until better means of control have been adopted. These consist either of placing a *tourniquet* or elastic bandage [275] on the limb above the wound as a temporary expedient, or of applying a *ligature* around the open mouth of the vessel as a permanent measure.

387. The *field tourniquet* consists of a pad to be placed over the artery and a strap and buckle to secure it in position.

The screw tourniquet.

This may be extemporized by a roller for a pad, fastened by a loop-bandage [438]; or by encircling the limb with a handkerchief over a smooth stone for a pad, and tightening the handkerchief by giving several turns to a short stick which has been slipped between it and the skin on the opposite side of the limb to the stone. The latter was the original of the *screw tourniquet* that is yet in use, and which consists of an encircling band that may be tightened by the separation of two metal plates consequent upon the turning of a screw. The screw and plates are usually placed over the artery, with a large roller beneath the lower plate, to give a trustworthy pressure along the arterial track.

388. But the tourniquet is merely a provisional arrest, as during transportation to hospital or while awaiting the arrival of expert assistance or the needful instruments. Permanent closure of the bleeding vessel must be effected by the *ligature;* and when an artery of some size is divided it is not enough to tie its upper or bleeding end—its lower end should also be secured to prevent hemorrhage from it after the establishment of the collateral circulation [187]. When an artery is tied firmly with a small round cord of ligature silk or carbolized catgut, its inner coats are cut across at the point of stricture and retract on account of their contractility, leaving only the tough external tunic in the grasp of the ligature. In the course of one or more

Ligation of an artery: The ligature has cut the middle and internal coats which are retracted on each side within the arterial tube; a conical plug of coagulum has been formed on the side towards the heart.

days, depending on the quantity of tissue devitalized by the pressure of the ligature, the whole of this tissue becomes absorbed and the loop is free to be withdrawn by traction; but this does not usually happen until the cellular coat on the upper side of the ligature has been sealed by exuded lymph and the canal of the vessel closed for some distance by a fibrinous coagulum. There is, therefore, no recurrence of the bleeding when the ligature is withdrawn as in the case of silk, or when it is liquefied and absorbed as in the case of catgut. Ordinary catgut is too readily softened and absorbed when imbedded in the living tissues; but when specially prepared, as by soaking for a long time in carbolized oil, it becomes harder, and resists the action of the tissues until time has been afforded for the consolidation of the artery.

389. In operating, the bleeding mouth of the vessel is transfixed with a tenaculum or seized with a spring or artery forceps [273], and drawn out from the mass of tissue in which it is imbedded, after which an assistant passes the ligature around it and ties it firmly in a reef knot or that modification of a reef knot sometimes called the surgeon's knot. In tying the reef knot, the middle of the ligature is applied to the artery immediately in advance of the points of the forceps or curve of the tenaculum, and its free ends are brought loosely around on the other side of the artery and crossed from one hand to the other, forming a ring through which one end is passed to form a single knot. This knot is drawn tightly on the artery, the traction on the

The reef knot.   The surgeon's knot.

ligature on both sides being made as close to the vessel as possible in order not to drag it out of its position. The inner coats give way under the pressure of the tightened ligature; but as the single knot would speedily slip, it must be guarded by a second, made as before by crossing the ends to form a ring, slipping one end through the ring and drawing on the ends; but in crossing the ends for this second single knot, that which lies in front of the first knot must be placed in front of the other, else the second knot will not hold securely. Sometimes, in tying the second, the first knot will become relaxed, to prevent which the tip of the finger of an assistant should be pressed upon it until the second is about to be run home; but the first may be made to hold better until the second is tied by passing the end through the ring twice, constituting the surgeon's knot, instead of once, as in casting the sailor's or reef knot. If the ligature is of catgut, both ends are cut off short at the knot, which is left in the wound to be absorbed; if of silk, one of the ends is cut short off and the other is brought out at the lowest part of the wound, so that it may act as a drain for the passage of liquids exuded after the lips of the wound have been brought into position and sutured.

390. The hemorrhage which takes place from a wound at the time of its infliction and ceases spontaneously or with the aid of surgical art, as described above, is called *primary*. This hemorrhage, whether arterial, venous, or capillary, must always be completely stopped before a wound is closed, lest it should continue and become infiltrated between the muscles or penetrate into neighboring cavities, forming a clot which would give rise to extensive inflammation by its subsequent disintegration and decomposition. The greater the shock attending the injury, or the greater the faintness from blood already lost, the greater must be the care given to this point, because, as the patient rallies from the shock or syncope and the impulse of the heart becomes stronger, there is greater danger of a recurrence of the bleeding. It is well, therefore, when there is much nervous prostration, to wait for marked indications of returning strength before finally closing the wound.

391. The bleeding which occurs during reaction, when the clots which have formed in the mouths of the vessels are driven out by the increased force of the circulation, is called *intermediary hemorrhage*. When care has been taken in the first instance, this seldom amounts to more than a mere oozing which may stain the overlying dressings.

392. But danger from hemorrhage has by no means ceased when reaction has been established, for at the end of several days the establishment of a collateral circulation [187] may give rise to bleeding if the lower end of the

divided artery was left unguarded by a ligature [388]; ulceration or sloughing may lay open an artery or vein the coats of which have been contused; or a ligature may come away, leaving the vessel imperfectly guarded by coagulum and exuded fibrin, so that what is called *secondary hemorrhage* may occur, and be even more dangerous than the primary bleeding, on account of the greater difficulty of securing an injured vessel in a wound that has been more or less altered by adhesions and granulations than in one that is recently inflicted. Secondary hemorrhage, if slight, may be controlled by pressure or the injection of styptic solutions; but if profuse it must be temporarily restrained by the application of a tourniquet or other pressure to the main trunk above the wound until competent surgical assistance has been obtained to open the wound and secure the bleeding vessels.

CHAPTER VI.

WOUNDS OF THE HEAD, NECK, AND TRUNK.

393. The gravity of WOUNDS OF THE HEAD depends on the amount of injury sustained by the nervous masses. Contused wounds should be treated, as in other parts of the body, to moderate inflammation and promote the absorption of extravasated blood or exuded inflammatory products. Incised wounds sometimes bleed freely, and require a compress. Plaster strips, not sutures, should be used to retain the edges in apposition, even when the wound is lacerated and the scalp torn from the underlying bone. Preliminary to this, the hair must be removed by close cropping and shaving.

394. Dressings are retained on the scalp in various ways. In applying a *single roller* or bandage for this purpose, the end is placed on the forehead and one or two firm horizontal turns are made, ending on the temple, where the superimposed layers are fastened together by a pin or stitches; the bandage is then carried from this fixed point over the vertex and under the chin as often as may be required, and the end and crossings stitched for greater security. The *double roller* is made by stitching together the free ends of two single rollers. In applying it to the head, horizontal turns are made to fix in their place those turns that are carried over the cranium to cover its dome. This, the *recurrent* or *capelline* bandage, is a relic of the olden time, when to do a thing *secundem artem* was considered of more importance than to do it so that it should fulfil the indications in the case under treatment. A simpler retentive appliance is the *four-tailed bandage.* This is a piece of muslin a yard long and six or eight inches wide which has been torn lengthwise into two strips of equal width, except in about

Single roller for the head.

Four-tailed bandage on chin.

six inches of its middle part. The untorn portion is applied to the vertex; the posterior tails are tied under the chin and the anterior tails under the occiput.

395. A narrower four-tailed bandage than that for the head is useful on the chin; the upper tails are tied on the back of the neck, or crossed there and tied under the chin, and the lower ones are carried to the vertex. The *six-tailed bandage* for the head is made on the same principle. The central part is applied to the vertex, the middle tails are made fast under the chin, the anterior tails beneath the occiput, and the posterior ends on the forehead. But a handkerchief used as a *triangular bandage* is the simplest and often the best form of retentive appliance. The centre of the base or long side of the triangle is applied to the forehead, the apex or opposite corner falling over the occiput, where the ends of the base meet and are tied firmly

together, or are brought forward after crossing, and made fast on the forehead.

396. In cases of CUT-THROAT there is often much bleeding, which should be controlled by continued compression with the finger if the bleeding is venous, by torsion or the ligature if arterial. When the cut is below the larynx, the head, neck, and chest may become swollen from the escape of air into the cellular tissue (*emphysema*); this is not usually a dangerous complication. The wound should be closed by many stitches, and its lips supported by long strips of plaster, stretching downward and outward toward the collar-bone and shoulder; but if the patient is unable to breathe through the natural passage the wound should be left open, with its sides approximated by a forward inclination of the head. When the pharynx is involved, nourishment must be administered by means of the stomach-pump [502].

397. WOUNDS OF THE CHEST penetrating the cavity and injuring the lung are usually attended with shock; and when there is serious internal hemorrhage the patient sinks into a condition of great prostration and final collapse; but however desirable it may be to ascertain the full extent of the injury, the finger or probe should not be used in the examination. *Penetration of the wall* is usually associated with the entrance of air through the wound, distending the cavity and preventing the filling of the lung during inspiration. Blood in the pleural cavity, *hæmothorax*, gives a dull sound on percussion over the collected extravasation; air in the pleural cavity, *pneumothorax*, gives a tympanitic or drum-like sound over the intruded air When the *lung is penetrated*, blood appears in the sputa, and an admixture of air gives a frothiness to liquids that escape from the wound; the cellular tissue of the chest often becomes infiltrated with air, constituting the condition called *emphysema*. If the penetration is small, as a stab, and inflicted with a clean instrument, it may heal as readily as a simple non-penetrating wound, even although there may be extravasated blood within the pleura and an injury to the lung; but if the wound become infected by injudicious handling, by the exposure incident to its open or gaping character, or by the lodgment of any foreign matter, as a bullet or fragment of clothing, a violent inflammation of the pleura and lung will result.

398. The treatment of the non-penetrating wound is that of flesh wounds in any other situation; and that of penetrating wounds differs only in the addition of precautions to restrain internal hemorrhage and moderate the subsequent inflammation. During collapse, stimulation must be attempted with caution, lest it aggravate the hemorrhage; during reaction, perfect rest and quietude must be enjoined.

399. WOUNDS OF THE ABDOMEN which do *not penetrate* into the abdominal cavity require greater care in their management than corresponding wounds of the chest. This arises from the greater likelihood of foreign bodies escaping detection among the muscular tissues, the greater difficulty of obtaining coaptation of the deeper parts of the wound, the greater danger of extravasated blood burrowing between the muscles, and, corresponding with all this, the greater tendency of such wounds to suppurate and occasion constitutional disturbance. In treating them, foreign matter should be carefully removed; bleeding should be entirely stopped before the edges are approximated, and this should be effected by the ligature if the flow is copious, or by torsion, styptics, cold or local pressure if slight; on no account should the wound be closed and compresses applied with the view of controlling hemorrhage, lest the blood find

its way into the intermuscular spaces; compresses should be used, however, to insure coaptation of the deeper parts, and the edges should be closed by sutures or strips of plaster while the body is retained in that position which most facilitates closure. In *punctured* wounds it is often difficult to decide whether the cavity has been penetrated; in such cases, exploration by the finger or probe is not advisable. Hemorrhage must be stopped before the wound is closed, as otherwise blood, if the wound penetrated, would collect in the peritoneal cavity. When *penetration* is evidenced by protrusion of the viscera, the case must be brought immediately to the attention of the surgeon.

CHAPTER VII.

THE CAUSATIVE CONDITIONS OF INSENSIBILITY.

400. Insensibility may arise from: Injury to the brain by external violence; apoplectic conditions, or those connected with congestion of the brain; and syncopic conditions, or those connected with a diminished circulation of blood.

401. CONCUSSION OF THE BRAIN varies in intensity from a momentary stunning, which leaves the patient temporarily confused, weak, and tremulous, to a state of continued insensibility with feeble but easy breathing, an almost imperceptible pulse, pale countenance, and cold, clammy skin. The occurrence of vomiting in a case of severe concussion is accounted a favorable sign, as the patient usually rallies after it. The patient should be put to bed and be warmly covered; warmth should also be applied to the extremities; and should the power of swallowing be retained, the removal of intense depression may be cautiously attempted by small doses, twenty or thirty drops, of aromatic spirit of ammonia. A purgative enema should be administered; and if insensibility continue for some hours, the catheter should be used. When reaction takes place, cold cloths or evaporating lotions should be applied to the shaven scalp, with perfect rest, quiet, an unstimulating diet, and an occasional laxative. If serum be effused or lymph exuded during reaction in a case of concussion of the brain, these morbid products, confined within the bony case, give rise to symptoms of COMPRESSION. If a blow on the head lacerate some of the blood-vessels within the skull, the resulting hemorrhage occasions a gradually increasing pressure on some part of the brain; or if the blow fracture the skull and drive a portion of the bone inward, the brain suffers immediately from the pressure of the intruding fragment. The symptoms and treatment of compression are those of *apoplexy*, for the latter is caused by the pressure of a hemorrhage or effusion which depends on disease instead of on external violence.

402. In CONGESTION OF THE BRAIN the arteries bring an excess of red blood and the capillaries become engorged. The symptoms are headache, fever, great restlessness, and delirium; and not until effusion has taken place and exercises a compression on the brain does the delirium subside into the insensibility and profound stupor of the apoplectic state. The gradual progress of the case indicates, in this instance, the conditions to which the insensibility must be attributed.

403. APOPLEXY. — Insensibility, as a direct result of active congestion, is gradual in its development; but as an indirect result it often occurs suddenly, particularly in elderly persons, in whom the walls of the cerebral arteries have undergone weakening and dilatation. The strain on these weak points during a congestion of even a trivial character may be such as to occasion their rupture and a sudden extravasation of blood into the substance of the brain. Pressure is exercised on the brain by this extravasated blood, and is manifested by the same symptoms as the pressure from an indriven portion of a fractured skull or the compression arising from blood extravasated as a result of violence. The patient is unconscious, his face flushed, pupils insensible to light and perhaps of different sizes, pulse slow, hard, and full, breathing slow and somewhat irregular, the inspiration snoring and the expiration puffing; generally the features are drawn to one side, indicating the existence

of paralysis [291]. The treatment consists in loosening the clothing about the neck and chest, and applying cold to the head, which, with the shoulders, should be kept elevated by pillows; hot water with mustard to the feet and legs, and the placing of a drop or two of croton oil on the tongue, with a purgative enema if the bowels are not opened in the course of an hour. Cloths or other absorbing material should be adjusted under the patient with a rubber sheet to protect the mattress; and if the insensibility has lasted for some time, a catheter should be introduced to evacuate the bladder.

404. SUNSTROKE is the result of heat, over-exertion, and an insufficient supply of water. If the water needful in the system be not replenished from time to time as it is dissipated in vapor from the lungs and skin, evaporation will cease; the skin, instead of being bedewed with moisture, will become dry; and the blood, altered by its loss of water and its excess of heat, will tend to stagnate in certain organs, as the lungs and brain. At first the patient may have merely a sense of oppression in the chest, or of headache, with increasing confusion of thought; but his limbs will speedily begin to tremble, and ultimately sink under him as he falls insensible. His pulse is quick, compressible, and small, sometimes intermitting; his breathing short, shallow, and interrupted by deep-drawn sighs; his lips are livid; he clutches spasmodically at his chest, as if to remove something that oppresses; and he may be seized every two or three minutes with violent convulsions which stretch him out stiff and rigid for a few seconds, and then gradually relax, with hesitating twitches which forebode a recurrence. A man in this condition will die very shortly; or he will recover consciousness in a short time, if properly treated, becoming perfectly well in a few hours or days; or some accident may occur to the brain during its congested condition, particularly if he be elderly and the subject of degeneration of the walls of the cerebral arteries, and in consequence of that accident he may remain unconscious for many days, and have paralysis or other indications of injury to the nervous system.

405. Manifestly, the sooner a sunstruck patient is treated and recovered from his unconscious and convulsed condition the less will be the risk of injury to the brain. He should be immediately carried to the nearest shade, where his outer clothing should be removed. If he still retain the power of swallowing he may take as much water—not iced water—as he will drink; but if he is so insensible that no effort is made at swallowing when a few drops of water are placed in his mouth, his surface should be assiduously mopped with dripping wet cloths, the moisture of which, while cooling by its evaporation, will be absorbed by the vessels of the skin, and will tend to relieve the abnormal condition of his circulation, so that in a short time the power of swallowing may be recovered. If, however, the patient fall into a comatose state, he must then be treated as a case of apoplexy [403].

406. Iced water should not be given to a person suffering from heatstroke. The cooling produced by it is merely local. It cools the fauces and the stomach, and is in small quantities grateful to the palate; but its influence in lessening the temperature of the body generally is no greater than that of the same quantity of water at the ordinary summer temperature. Ingested water operates during its exhalation from the lungs and skin, but to get there it must first be absorbed into the circulation by the minute vessels of the stomach. Cold contracts these vessels and interferes with absorption. A person in ordinary health who desires to experience the feeling of local cooling may take a glass of iced water, as he can afford to wait for its absorption until the subsidence of the temporary chill; but one who is dying for want of water should by no means have the absorbing powers of his stomach rendered inactive by the temperature of the water supplied to him.

407. Epileptic Stupor.—An attack of epilepsy consists of loss of consciousness and convulsions. It varies in severity from a momentary lapse of mind, with barely appreciable spasms, or twitchings of the features, to a violent paroxysm lasting two or three minutes. When severe the patient becomes suddenly pale, falls forward on his face, sometimes with a loud, quick scream, and is immediately seized with a rigid spasm which lasts for ten or twenty seconds, and is followed by rapid contractions and relaxations of the muscles, which contort the features and jerk the body and limbs with much violence. These last for a minute or two, during which the face becomes congested, the pupils dilated, and the lips covered with foam, which may be tinged with blood if the tongue has been injured by the spasmodic action of the lower jaw; there may be also involuntary discharges from the bowels and bladder. A long-drawn sigh ends the fit, and the patient sinks into a stupor which may last for several hours and be mistaken for apoplexy on account of its heavy breathing and puffing expirations. The history of the case is, however, distinctive: The patient is perhaps recognized as an epileptic subject, or his present condition of stupor may be known to have been preceded by convulsions; the presence of foam on his lips is suggestive of the epileptic struggle, and the absence of paralysis tends to confirm this diagnosis by excluding apoplexy as a cause.

408. In treating an epileptic a handkerchief or a gag of wood should be placed in the mouth of the patient to protect his tongue; the neck and chest should be freed from all constrictions; and no more restraint should be used than is required to guard against accidents. After the fit he should be placed in bed to recover from its exhausting effects.

409. Convulsions, with temporary loss of consciousness, occurring *in young children* who are not known to have disease of the brain, are usually due to the irritation of teething or to some disorder of the digestive organs, as the presence of indigestible matters or worms. The patient should be placed across the lap of a nurse, and while ice-cold water is applied to the head, the feet and legs should be immersed in a pail of hot water. Meanwhile the gums should be examined; and if a rising tooth is found to be the occasion of much redness and swelling, a free cut should be made over it, lengthwise of the gum, to relieve the pressure. If there is no tumefaction of the gum, an enema of soap-suds should be administered, or an emetic of ipecacuanha, if there be a history of some indigestible substances taken into the stomach.

410. Alcoholic stupor is distinguished from apoplexy mainly by the condition of the pulse and pupils—the former is rapid, small, and soft, the latter dilated; moreover, there is no contraction of the facial muscles or other indication of paralysis; and cold water freely applied to the head has usually some influence in arousing the patient from his lethargy. In an unknown case the odor of alcoholic stimulants in the breath should not be permitted to have weight as evidence, because apoplexy or brain injury is by no means inconsistent with the presence of alcohol in the system. In every case of doubtful causation, the patient should be taken into hospital and treated as if suffering from congestion of the brain until his actual condition is unmistakably revealed. Even when his stupor is known to have been occasioned by his own vicious habits, he should be treated as sick until the recovery of his intelligence. To confine a man in this comatose condition in a guard-cell merely exposes him to danger; it is no punishment, for punishment can be inflicted only when the individual is sensible of its application.

411. The treatment consists of cold to the head and an emetic of mustard, if the patient have not already vomited freely. If the pulse become weak and the skin cold and clammy, hot coffee should be administered, or small doses of aromatic spirit of ammonia in water, with frictions and warmth to the general surface.

412. DELIRIUM TREMENS.—Prolonged alcoholic excess leads to a condition of nervous prostration characterized by muscular tremors and delirium. The patient is constantly anticipating some fancied evil or annoyance, which keeps him in a state of nervous inquietude, talking anxiously and with more or less incoherence. He is usually easily controlled and diverted from his intentions, but he must be watched closely for fear of accidents. Milk and beef essence should be given as food, with bromid of potassium or chloral, or morphine by hypodermic injection, to quiet the nervous system and induce sleep. For OPIUM NARCOTISM, see par. 523.

413. INSENSIBILITY FROM COLD.—Exposure to extreme cold acts like opium in exercising a benumbing influence on the sensations, leading from drowsiness to a stupor so profound that the system fails to recognize the necessity for breathing. If the body is stiff, the tissues should be relaxed by friction with cold water, after which artificial respiration

offers the only prospect of resuscitation. Even when the insensibility has merely embarrassed but not arrested the respiratory movements, the return of warmth to the body must be gradually effected, lest the patient die in the delirium of excessive reaction. Only such stimulants as hot coffee, beef tea, and aromatic spirit of ammonia are admissible.

414. Cases of insensibility associated with cerebral congestion, and arising from causes which interfere with the respiratory function, such as *strangling, choking, smothering, drowning*, etc., are referred to under the heading of Artificial Respiration, page 241.

415. HEAT EXHAUSTION.—Insensibility from exhaustion is a common occurrence on forced marches, particularly among raw troops. Ordinarily the order of march provides for a rest at the end of every hour to refresh the men, to permit those who have fallen behind to recover their position in the ranks, to readjust accoutrements, and to replenish water supplies when occasion offers; but on a forced march, those who have fallen behind have no opportunity afforded them of rejoining their company. They struggle on through the heat and dust until they become utterly exhausted and drop unconsciously by the roadside, face pale, lips bloodless, pulse rapid and feeble, respiration sighing, and the muscular system affected with the tremors of prostration. Recovery in these cases is speedy when the conditions are favorable. The removal of the patient's belts and burdens, rest in the recumbent position in the shade, cold water to the head and face, upward frictions to the limbs, and stimulants and water as required, generally suffice to re-establish his powers. Frequently, however, these cases of heat exhaustion are complicated with conges-

tive tendencies, manifested by dryness of skin, pulmonary oppression, headache, and convulsive seizures [404]. In such instances alcohol is contra-indicated; water alone, or with a few drops of aromatic spirit of ammonia, is the proper remedy.

CHAPTER VIII.

ARTIFICIAL RESPIRATION.

416. A person at the point of death from asphyxia [227] may be recovered by keeping up movements of the chest that will alternately compress and expand the lungs; but no method of doing this is efficient unless *the tongue of the patient be prevented from blocking up the air passage.* During insensibility it tends to fall backward in the throat and close the upper end of the larynx. It must be seized, pulled forward, and held in that position by an elastic

Silvester's method: Inspiration.

band, string, or tape passed around its base and under the chin, by a pencil placed across its base and lodged in position behind the back teeth, or by the fingers of an assistant enveloped in a dry handkerchief to prevent them from slipping.

417. In SILVESTER'S METHOD the patient is placed on his back, with a pillow or roll of clothing under his shoulders to elevate them and carry the chin away from the breast-bone. The operator kneels at his head, and, taking hold of an elbow in each hand, makes the inspiratory movement by drawing the arms outward away from the chest, and then, continuing the motion, forcibly upward over the head of

Silvester's method: Expiration.

the patient. This drags the lower ribs upward and outward, expands the chest, and causes the entrance of air. The expiratory movement is made by passing the arms down again along the side and front of the chest, and making pressure on them when in that position, to force the ribs downward and drive the air out of the lungs. This completes the respiratory movement, which should be repeated at the rate of fifteen times per minute, until natural respiration returns or the case is abandoned.

418. In MARSHALL HALL'S METHOD the patient is placed

on the floor or ground with the face downward, his forehead resting on one arm, and a roll of clothing supporting his chest. While in this position the weight of the body compresses the ribs and expels the air from the chest—an artificial expiration which is deepened by making pressure on the lower ribs. Then the operator, with one hand on

Marshall Hall's method: Expiration.

Marshall Hall's method: Inspiration.

the patient's free arm near the shoulder, and the other placed under or in front of the corresponding hip-bone, rolls the body from face downward to its side and a little beyond. An assistant aids in this movement by handling the head and underlying arm. When the body has been thus rolled somewhat more than half round, the chest becomes relieved from superincumbent weight, and a certain volume of air enters. After resting a second or two in this attitude of inspiration, the patient is returned to the prone position, and pressure made along the ribs to imitate the expiratory act.

419. Other plans have been advocated, such as that of *compressing the abdominal walls* as if the object were to drive the contents of the abdomen upward into the chest. This forces the diaphragm upward and empties the lungs, after which the sudden withdrawal of the pressure permits of inspiration. Whichever method is used, it is well, during its progress, to attempt to stimulate the respiratory powers by holding ammonia to the nostrils and slapping the chest alternately with cloths wrung out of cold and hot water. As soon as natural breathing has been restored, efforts should be made to promote the circulation by rubbing the limbs in the direction of the trunk; to restore the warmth of the body by the use of warm clothing, blankets, hot flannel, bottles, or bricks, and to stimulate the vital actions by small doses of aromatic spirit of ammonia.

420. The conditions which call for artificial respiration are those in which the lungs are sound but deprived of air, either mechanically or by the substitution for air of some gas or vapor which is not directly poisonous, or when respiratory action has ceased on account of some narcotic influence exercised on the brain. These include:

(*a*) *Strangulation* by cord, ligature, or other compression on the neck or windpipe.

(*b*) *Smothering* from the mechanical closure of the nostrils and mouth by any substance, or even by interference with the movements of the chest, as when one has been imbedded to the neck in a sand-slide. In these cases the mechanical cause must at once be removed, together with articles of clothing which might interfere with the movements of the chest. If the mechanical cause has occasioned other injury than strangulation or smothering, artificial respiration may be of no value, as when in hanging the cervical vertebræ have been dislocated or fractured; but if there be no evidence of this, and particularly should there be a spasmodic quivering of the muscles or the faintest indication of pulsation of the heart, efforts should be made without loss of time to arterialize the venous blood by imitating the movements of respiration.

(*c*) *Choking* from the pressure of something in the gullet. When a man becomes suddenly suffocated while eating, the fingers should be immediately passed over the base of the tongue, and as far down as can be reached, to bring away any foreign substance that may be there [432]; after which artificial respiration is in order, if necessary.

(*d*) *Drowning* by immersion in water: No time should be lost in removing the body from the water; and except in inclement weather, resuscitation should be attempted on the spot. The upper or body clothing should be removed, and, as the air tubes are frequently choked with indrawn water, the patient should first be placed face downward on the ground, and his body raised by the hands of the operator clasped underneath the abdomen to cause the intruded water to escape, partly by drainage and partly by the upward pressure on the lungs occasioned by the constriction of the abdomen. The mouth and nostrils are then cleaned and artificial respiration is instituted. Meanwhile the lower garments should be removed, and the surface dried and protected with blankets or shawls.

(e) *Asphyxia from coal-gas, charcoal fumes, or the exhalations of vats, pits, or mines.* Care must be taken in recovering the body that the rescuer, if overcome by the harmful atmosphere, may be promptly withdrawn. The noxious gas in such cases consists mainly of carbonic acid, *carbon dioxid*, sometimes associated with carbonic oxid, *carbon monoxid*. The former suffocates by preventing aeration of the blood, while the latter has a directly poisonous influence; but as coal-gas contains less than ten per cent. of the poisonous oxid, there is a prospect of favorable results if treatment by artificial respiration and cold affusion be promptly instituted.

(f) *Insensibility from ether or chloroform* [276], *carbonic oxid, from opium* [523], *or from exposure to extreme cold* [413]. The escape of water-gas gives rise to most of the deaths from carbonic oxid; the gas contains thirty or forty per cent. of this insidious poison. When opium is the cause of danger, artificial respiration aims to prolong life until the poisonous influence has been eliminated or counteracted by other means.

421. Signs of Death.—There is no positive sign of death except the occurrence of decomposition in the tissues of the body. This is the statement of a scientific fact; but at the same time there is practically no difficulty in determining the presence of death long before putrefaction sets in. The evidence of death consists of the concurrent existence of several conditions: 1. The continued cessation of breathing, as determined by a polished surface held over the lips and nostrils; 2. The continued absence of any movement of the heart when closely watched for by the ear and touch; 3. Relaxion of the muscles, passing into rigidity at the end of three or four hours; 4. The glazing of the eyes; 5. The progressive cooling of the body from its normal living temperature to that of the surrounding atmosphere. Many single tests of the presence of death have been suggested. Perhaps the best of these is the tying of a string around the end of a finger or lobe of the ear: no change takes place on the dead body; but if there be any vital movement in the vessels the point beyond the ligature becomes more or less congested. In the *wrist test* a splint is placed in front to protect the radial and ulnar arteries; a cord is then tied tightly around the splint and wrist to block the veins on the back of the hand. These veins gradually become filled with blood if the heart's action and circulation have not wholly ceased. The hypodermic injection [502] of a few drops of a strong solution of ammonia has also been recommended. If the body be not dead the ammonia will produce on the skin, over the point where it was injected, a bright red patch, on the surface of which raised red spots will appear; but in absolute death a brown dark blotch will be developed.

CHAPTER IX.

FOREIGN BODIES IN THE EYE, EAR, ETC.

422. *Foreign bodies* lodging in the *eye* cause much redness and distress. If they are not visible on careful inspection of the globe, they should be looked for behind the upper eyelid. To do this a probe is laid along the lid from the root of the nose outward, and pressed lightly against the upper part of the ball; at the same time the lid is

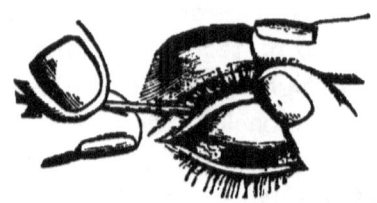

Everting the upper eyelid. The conjunctival vessels are congested.

pulled away from the globe by traction on the eyelashes, and the patient is told to look downward, when the lid will be tilted over the probe behind it, fully exposing its conjunctival surface. The intruding substance should be brushed away with a camel's-hair pencil or pledget of lint; if it be partly embedded in the mucous membrane it should be lifted out with the flat surface of a blunt-edged lancet. A drop of olive or castor oil instilled into the eye will relieve the irritation.

423. Eye-drops, such as solutions of nitrate of silver, atropin, etc., are not *dropped* into the eye, but *instilled* or insinuated. The patient turns his face upward, and the operator, having depressed and partly everted the lower lid with the fingers of one hand, brings a drop hanging on the end of a glass rod, tube, or dropper into gentle contact with the mucous membrane, from which it flushes over the surface of the eyeball.

424. When a particle of extraneous matter is embedded in the conjunctiva the blood-vessels become enlarged, forming a close-set network such as is represented in the illustration just submitted. The enlarged vessels give rise to a feeling as if sandy particles were lodged on the surface of the membrane. Usually the removal of the irritant cause, the seclusion of the patient in a shaded room, a dose of aperient medicine, and the use of cold lotions will speedily allay the congestive action; but if the redness and other symptoms persist after these measures have been tried for a day or two, a drop of a solution of nitrate of silver, one or two grains to the fluidounce of water, instilled morning and evening, will generally be efficient.

425. Sometimes, however, the eyelids become swollen and puffy, closing over the eyeball, yet allowing the tumefied, red, and inflamed mucous membrane to protrude between them and giving issue to a free discharge of purulent matter. As this matter is infectious, the greatest care must be used to prevent the spread of *purulent ophthalmia* by infected fingers, towels, etc. The disease is apparently due sometimes to atmospheric changes, in which case both eyes of the patient will suffer simultaneously; at other times it results from the infection of *gonorrheal discharges*, and in these cases one eye is usually first affected, and is seriously endangered before the other receives the contagion. Rest, quiet, low diet, and aperient medicines are required as constitutional remedial agents, whilst the local measures consist of repeated syringing and warm fomenta-

tions to clear away accumulated matter, scarification of the tumefied membrane if it project much through the fissures between the lids, and the introduction of a solution of nitrate of silver to stimulate the enlarged vessels to contract. Neglect in a case of this kind may lead to total loss of vision by *sloughing* of the *cornea*, as that part of the eye is dependent on the conjunctiva for its supply of nutriment.

426. When men have been exposed for a considerable time to a garish light, as that reflected from stretches of sand or snow, the retina or sensitive membrane of the eye may become temporarily strained and unable to take cognizance of ordinary impressions. This condition, which has received the name of *snow-blindness*, usually requires no other treatment than rest in the shade for a few days to enable the nervous matter to recover its normal functions. Sometimes, however, the local reaction which takes place becomes developed into a general inflammation of the eye, requiring special surgical attention.

427. *Foreign bodies in the nose.*—Young children sometimes push beans, buttons, or other small objects into the nose, and are unable afterward to withdraw them. They may sometimes be removed by a vigorous effort to blow through the blocked-up nostril or by sneezing artificially induced; failing this, they must be withdrawn by a fine forceps or blunt hook, afterward washing out the cavity with warm water in which a little common salt has been dissolved.

428. Sometimes, in southern climates, uncleanly children or intemperate adults become affected with *maggots* in the nasal cavities. These cause much local irritation, loss of sleep, fever, and ultimately delirium. They may be cleared out by making the patient inhale half a drachm of chloroform and immediately thereafter syringing out the cavities with a warm solution of common salt, repeating the opera-

tion at intervals of a few hours, until all the maggots have been removed.

429. *Foreign bodies in the ear.*—Insects cause great distress, and should be immediately dislodged by filling the canal with warm oil or gently syringing with warm water. The latter procedure generally suffices to remove small bodies, such as beads, seeds, etc.; if it fail, the case should be left for fully qualified surgical skill. Plugs of cotton-wool may be withdrawn by forceps after dissolving their waxy adhesions by instilling a few drops of olive oil. Wax should not be picked out, but softened by instillations of oil at night and syringing with warm water next morning.

430. The ear should never be syringed with cold water; the water used should always be distinctly warm to the fingers. The patient, or an assistant, holds the edge of a

Syringing the ear.

cup or receiver closely to the side of the neck below the lobule. The operator takes hold of the upper and back part of the ear with the fingers of one hand, pulling it upward, backward and slightly outward to straighten the canal, while he lays the nozzle of the syringe just within its entrance and pushes the piston gently and slowly home. Two or three two-ounce syringefuls generally suffice to cleanse the canal. After syringing, the water should be permitted to drain from the ear, which should be dried with the twisted corner of a towel or fragment of lint.

431. *Foreign bodies in the trachea.*—When buttons, beads, seeds, small coins, etc., enter the larynx or trachea of children, efforts at removal should be restricted to coughing while the patient is held with the head and body inclined downward.

432. The *foreign* bodies which stick in the *pharynx* are usually fish-bones. If the bone can be seen, or felt by the tip of the finger, when the tongue is depressed, it should be removed by the fingers or forceps. If it be too low down for this procedure, large mouthfuls of half-chewed bread should be swallowed to carry it along. It must be remembered that the tearing of the mucous membrane will for some time afterward give rise to the impression in the mind of the patient that the bone has not been dislodged. Large bodies, such as fragments of meat causing choking, must be immediately removed with the fingers.

433. *Foreign bodies in the stomach.*—When a child has swallowed any of the small bodies mentioned, the alarm of friends should be quieted and the patient fed for a day or two on oat-meal, puddings, etc., after which a laxative dose of castor-oil should be given, with instructions to look for the foreign body in the intestinal discharges.

CHAPTER X.

FRACTURES.

434. A fracture, surgically, is a break in the tissue of a bone. It is *simple* when the bone is broken at one point; *comminuted* when broken into more than two fragments; *impacted* when one fragment is wedged into the other; *compound* or *open* when a wound penetrates from the surface to the plane of the fracture; and *complicated* when it is associated with other serious injuries, as contusion, laceration, etc. A fracture may be comminuted and the soft parts contused and lacerated, yet if the skin be unbroken, extravasated blood and effused serum are absorbed and satisfactory repair goes on; but in a compound fracture which becomes infected by its exposure to the air, violent inflammation, burrowing suppurations, and sloughing cause much suffering and danger.

435. The symptoms of fracture are pain, swelling, loss of power, deformity, unnatural mobility, and crepitation. There is nothing characteristic in the first three, as they may be present in bruises and dislocations; but angular deformity in the case of a long bone manifests the nature of the injury as definitely as the mobility in a part that ought to be solid, or the *crepitus* or grating when the broken fragments are moved on each other. All fractures of the long bones are more or less oblique; and the shortening and difficulty of retaining the fragments in their proper relations to each other are proportioned to the obliquity; the lower fragment, as being the more movable, is the one displaced.

436. A fracture should be treated as soon as possible after the accident, because when swelling and muscular contraction have occurred it is more difficult to bring the broken ends into accurate contact (*coaptation*). When a bone lies so near the surface that the fingers can recognize it, it is easy to ascertain whether coaptation has been effected; but when the bone is well covered with muscles, the recognition of a proper setting is more difficult, and must depend upon the removal of all observable deformity.

437. *Splints* are used to keep the broken ends in contact until union is effected. They are of wood, pasteboard, perforated felt, gutta-percha, strong leather, sheet-tin, or zinc, or any other material which possesses the proper strength, and can be cut to suit the necessities of the case. The angles formed by the meeting of the sides and ends should, in all splints, be rounded off; the edges of metal splints smoothed, and of wooden ones bevelled. Pasteboard, leather, and gutta-percha must be softened in hot water before they can be moulded to the part for which they are intended. Gutta-percha requires practice for its successful application, inasmuch as a slight excess of heat transforms it into a sticky, unmanageable mass. It should be folded in muslin which extends beyond its ends, and immersed in hot water while held by the free ends of its muslin wrapper until the needful plasticity is attained. In the absence of anything better, a handful of selected straw may be cut to the proper length and quilted between the folds of soft cloth.

Reversing roller on forearm.

438. All splints, of whatever material, must be padded before being put to use over the fractured bones. Several layers of sheet wadding are cut to give the needful thickness of pad on the face of the splint. These are retained in position by being folded with the splint in muslin as a packet is folded in paper, the free ends being brought back and stitched together behind, leaving the front of the pad smooth and unwrinkled. Oakum, patent lint, and cotton batting are also used for this purpose, but they do not make so smooth a pad as the sheet cotton. When the bones have been brought into their proper place the splints are applied and retained in position by a bandage, or preferably by strap-and-buckle fastenings, because they leave the limb partly in sight, and can be relaxed or tightened with facility. The *loop bandage* answers in the absence of straps; a length of about three feet is cut from a roller, and doubled lengthwise on itself; this is passed around the splinted limb and the free ends are drawn tightly through the bight; these ends are then separated from each other and tied firmly at their meeting on the opposite side of the limb.

439. In applying a *roller* for the retention of splints or any other purpose, each turn is made to overlap that which preceded it; and when the increasing thickness of the limb prevents the roller from lying flat and at the same time preserving the overlap, the difficulty is overcome by what is called *reversing*. The upper border of the turn is folded obliquely over the forefinger of the left hand, while the roller in the right hand is made to describe a half turn on its long axis, after which the circular movement is continued until the point is reached where the next reverse has to be made. If a roller which encases or encircles a limb is likely to become wetted, it should be applied lightly, else the contraction which follows the wetting will make it harmfully tight.

440. Sometimes the bandage itself is made to act the part of a splint by saturating it with some substance, such as starch, soluble glass, or plaster of Paris, which will endue it with a solidity and resistance sufficient for the purpose. Whichever of these materials is employed, the limb is first lightly encased in an ordinary roller, and its hollows and inequalities filled with cotton batting, after which the so-called *immovable dressing* is applied. A colored tape should be placed longitudinally beneath this dressing to act as a guide for the knife, scissors, or cutting pliers used in its

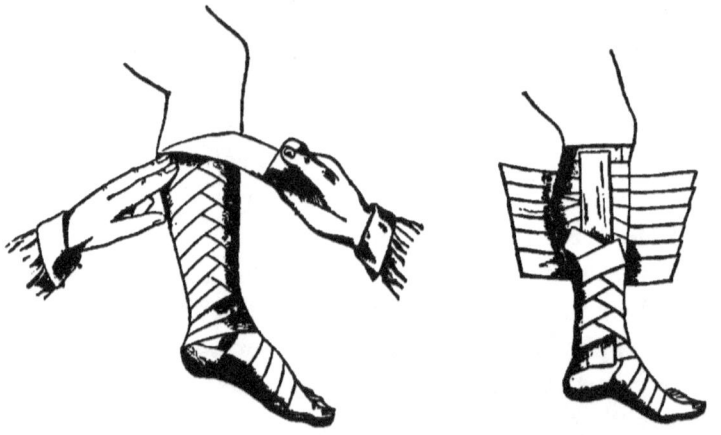

Reversing roller on leg.　　Many-tailed bandage over starched splints.

subsequent removal. The unyielding character of these dressings renders them applicable only in fractures in which there is no danger of the development of inflammatory swelling.

441. *Starch* is prepared as for laundry use. The bandages to be used are unrolled, soaked in the starch solution, re-rolled, and applied as an ordinary roller. Strips of saturated bandage may be laid on longitudinally to strengthen the casing, after which another circular or spiral starched roller is used, or a *many-tailed bandage* may be employed. This latter consists of strips cut long enough to go once and a half round the limb. These are arranged on a towel or pillow, the upper one being laid down first, and each successive strip overlying that which preceded it by half its width. The strips are then brought beneath the limb, around which they are applied as shown in the cut. The starch bandage takes from twenty-four to thirty-six hours to set. Should it become loose by the subsidence of any swelling existing at the time of its application, it must be cut along the line of colored tape, its edges trimmed, and the casing again closed and sustained in position by an ordinary roller.

442. A saturated solution of the *silicate of soda* may be used in the same manner as the starch solution. It sets slowly, but makes a light and firm support.

443. *Plaster of Paris* is of use as a support in fracture by virtue of its setting readily when mixed with an equal volume of water. Usually when thus mixed it consolidates in three minutes; common salt hastens, mucilage or glue retards, the consolidation; white of egg beaten up with the plaster gives it greater tenacity. To form a splint of this material, several thicknesses of washed muslin are folded to the required shape; these are immersed in and saturated with the liquid plaster, refolded to the proper shape, smoothed, and applied to the part, which is held in position until the plaster sets. When a limb is to be encased the dry plaster is first rubbed into the bandages which are to be used, and the limb is wrapped in cotton wadding, which is kept in place by an ordinary bandage; after this the prepared rollers are soaked in water and applied in circular turns without reverses; some of them may be cut into lengths and laid on longitudinally to give greater strength to the casing.

444. Union between the broken ends of a bone does not usually begin to take place until about ten days after the injury, so that any malposition detected during this period is susceptible of rectification. Three to six weeks are required for a firm cementation of the ends, the time depending on the size of the injured bone and the accuracy of the coaptation; small bones, as those of the fingers, unite more readily than larger ones.

SPECIAL FRACTURES.

445. Fracture of the skull may occur without notable symptoms other than those of the concussion [401] which might be present irrespective of the fracture; and they may get well without at any time giving symptomatic evidence of their existence. Sometimes a fissure or depression may be felt through the scalp, and, when the base is the seat of injury, there is occasionally a watery discharge from the ears, a continued bleeding from them, or the appearance of blood extravasated beneath the conjunctiva. As soon as the patient rallies from the concussion, his head should be shaved and cold applied, purgatives administered, and rest and quiet enjoined. Should symptoms of compression be present *from the moment of the injury*, there is probably some displacement of bone; should compression be developed with increasing intensity *some time after the injury*, it probably results from hemorrhage within the cranium.

446. When the fracture is *compound*, its edges may be discovered by exploration with the finger; and if any part of the bone is found to be driven inward, a cautious attempt is made by the surgeon to remove the fragment or raise it to its proper level. Slight symptoms of compression may sometimes be removed by relieving the brain of this pressure; but in these fractures the compression, when severe, is usually caused by an associated hemorrhage, and is unaffected by the restoration of the bone.

447. When the brain is laid open, foreign matter should be gently removed and the wound cleaned with an antiseptic solution, after which a compress should be laid over it and retained by a bandage to prevent protrusion of the brain substance.

448. The arch is the most common site of FRACTURE OF THE LOWER JAW. Crepitus and irregularity in the line of the teeth readily indicate the nature of the injury. A pasteboard splint is moulded to the front and under surface of the jaw, and bound in position by a *four-tailed* bandage [395]. The splint below and the teeth of the upper jaw above preserve the proper relations between the fragments.

449. The SPINE is fractured by falls from a great height or the fall of a heavy weight on the body when bent. The injury is usually attended with profound shock. Its symptoms depend on the situation of the fracture and the pressure on, or damage to, the spinal cord. Local violence may break off one or more of the *spinous processes* that project behind, without much injury to the contents of the spinal canal; but when the *bodies* of the vertebræ are involved in the fracture, there is usually paralysis of those parts of the body which receive their nerves from the cord below the seat of injury. Any necessary movement of the patient should be made with the utmost care, to avoid increasing the injury to the cord or its membranes. To turn him over on his face for a thorough investigation of the fracture would invite such a displacement of the lower portion as might cut the cord in two.

450. In these fractures there is severe pain, with swelling and other signs of contusion, irregularity of the spinous processes, and possibly crepitus on movement. When the cord is injured in the loins, the lower extremities are paralyzed, the urine retained, and the fæces passed involuntarily; when in the dorsal region, the symptoms include, in addition, a tympanitic swelling of the abdomen, due to paralysis of the muscular coat of the intestine; and when in the neck, the upper extremities become paralyzed. Fractures of the spine have generally an unfavorable ending, reached sooner or later according as the site of the injury is in the upper or lower part of the column. Above the third cervical vertebra death may be immediate, from injury to the phrenic nerves, which preside over the respiratory movements of the diaphragm; in the lower part of the cervical region, the patient lives only a few days; in the dorsal region, he may linger for two or three weeks, and in the lumbar region for one or two months, greatly distressed toward the end by increasing bed-sores. The patient should be put to bed in the position which gives him least discomfort, and the catheter must be used regularly to prevent distention of the bladder.

451. The RIBS are frequently fractured by blows, falls, or the counter-pressure of opposing forces, as when a wheel passes over the chest. The middle ribs are most liable to injury, as the upper have a better protection from the overlying soft parts, and the lower have a greater elasticity. Fracture is detected by pressing the finger along the skin over the line of the rib supposed to be injured, or by placing the hand on the part while the patient makes an effort at coughing. It is treated by confining the chest in a bandage—either several turns of a broad roller, or a closely fitting jacket fastened by straps and buckles or eyelets and laces—to keep the parts at rest and throw the burden of the respiratory movements on the abdominal muscles.

452. When the COLLAR-BONE is fractured near its middle, the inner fragment remains in its place, but the outer is dragged downward, forward, and inward by the weight of the extremity. A wedge-shaped pad is fitted, base upward, into the armpit; the arm is brought slightly inward across the chest, and the elbow and humerus are pressed upward so as to force the point of the shoulder upward and backward. A long roller is then applied to retain the limb in this position. A few turns are passed around the chest to fix the end, after which the arm is included with the chest in each turn; the roller is then carried below the elbow and over the opposite shoulder, and the operation finished by some circular turns to fix the whole. The various turns should be stitched here and there to prevent displacement.

453. When this bone is fractured near its outer extremity, there is little tendency to displacement, and all that is needful is to keep the arm bound to the side and to support the forearm in a sling.

454. The flatness of the SHOULDER-BLADE preserves it from fracture, but occasionally the *acromion process* is broken off. The patient complains of a great sense of weight, and the roundness of the shoulder is lost; but the deformity disappears when the arm is pushed perpendicularly upward so as to raise the broken fragment. In treating this injury, a cushion is placed between the elbow and the side to relax the deltoid, and the arm is kept at rest in this position by being bandaged to the side and having the elbow and forearm supported in a short sling.

455. A transverse fracture of the HUMERUS above the condyles looks like a dislocation backward of the radius and ulna at the elbow [481], because there is an unnatural prominence behind the joint, with a depression above it, and the front of the forearm is shortened; but it differs from a dislocation in that if the arm be fixed and traction made on the forearm, the unnatural appearance will be removed and the grating of the fractured ends may be perceived. Treatment: Reduce by traction, and preserve in position by a roller bandage applied around the lower part of the arm and the upper part of the forearm; use two padded splints—one straight, in front of the humerus, the other bent at a right angle, with one part behind the humerus and the other below the forearm. Fasten by buckle or loop bandages, and support the elbow and forearm in a sling.

456. Sometimes from the nature of the violence the *outer* or *inner condyle* is broken off, causing a prominence which may simulate dislocation. Crepitation may be felt when the inner condyle is fractured by grasping the prominence and at the same time moving the forearm backward and forward; when the outer condyle is fractured, crepitation is best elicited by alternately pronating and supinating the hand [157]. In treating these fractures, keep the forearm at a right angle with the arm, the back of the hand turned upward and the fingers bent if the inner condyle is involved, and the palm upward and the fingers straight if the outer condyle is the subject; then apply a roller around the lower part of the arm and upper part of the forearm, and over this a padded rectangular splint of pasteboard or other suitable material, placing one part behind the humerus and the other below the forearm, and securing all with a roller. The injured limb should be supported in a sling.

457. When swelling comes on quickly in fractures near the elbow, it is well to support and rest the limb, and to apply evaporating lotions to subdue inflammation before splinting the injured bone.

FRACTURE OF THE RADIUS. 263

458. When the *shaft of the bone* is broken, there is no difficulty in recognizing the injury. The inability of the patient to raise his arm, its mobility when handled, crepitation, and angular deformity sufficiently announce its nature. When the fracture is in the *lower part* a splint should be applied in front and another behind; but if the fracture is *about the middle*, the splints should be placed on the inner and outer sides. When the fracture is in *the upper part*, the upper fragment may be dragged outward and the lower fragment inward, or vice versa, according as certain muscles are broken off with the one or left attached to the other [154]. In either case traction should be made on the lower fragment until all shortening of the limb is removed, when three splints should be applied, one along the front, a second behind, and the third on the outer aspect of the arm. A wedge-shaped pad is then placed along the inner aspect, with its thick end in the armpit if the upper fragment be displaced inward, so that when the arm is afterward bandaged to the side the pressure of this pad may keep it in its place; but if the lower fragment be displaced inward, the thick end of the pad should be below and its apex in the armpit. In bandaging the arm to the side the roller should be applied firmly over the fragment which has the outward tendency and lightly over that which tends inward.

459. The weight of the arm in all these fractures of the humerus should be supported by a sling; but care should be taken that it does not press the elbow upward, lest it give rise to shortening of the bone in cases in which the line of fracture is oblique.

460. Fractures of the RADIUS are characterized by loss of the power of voluntarily effecting the movements of pronation and supination; the hand is supported by the patient

in the prone position. When the fracture is in the middle or lower third, it may be felt by the finger; if it be in the upper third, where the bone lies deeper in the muscles, it may be detected by grasping the upper part of the forearm in one hand, pressing the thumb firmly on the head of the radius, on the outer aspect of the posterior prominence of the elbow, and alternately supinating and pronating the injured limb with the other hand. The fracture is revealed by the grating of the fractured ends on each other and by

Fracture at wrist.

the failure of the head of the bone to participate in the rolling motion given to the lower fragment. When the fracture is near the wrist—Colles's fracture—its site becomes swollen by effusion into the sheaths of the tendons, which gives the injury the appearance of a dislocation, the more so that this same swelling interferes with the motion of the fingers; but crepitus readily distinguishes between the fracture and the dislocation.

461. Fracture of the ULNA may usually be discovered by drawing the finger along the edge of the bone on the posterior and inner aspect of the forearm. There is a depression at the seat of the fracture on account of the lower fragment settling toward the radius.

462. Fracture of *both bones* is readily recognized by the unnatural mobility, the angular deformity, and the grating of the fragments.

463. In splinting these fractures, the forearm should be bent at a right angle to the arm to equally relax the mus-

cles on the anterior and posterior aspects, and the hand should be kept midway between pronation and supination, *i.e.*, with the thumb upward and the little finger downward, because, if union were to take place with the hand prone, the power of supination would be restricted, and if with the hand supine, pronation would be impaired. If both bones are broken, the hand should be kept in line with the forearm; if the radius alone is fractured, the hand should be slightly depressed; and if the ulna alone is injured, the hand should be raised. Two splints are required, one for the back and one for the front of the forearm, the former reaching only to the wrist, the latter to the roots of the fingers. These splints should be broader than the breadth of the forearm, that the bandages may not compress the bones laterally; and they should have a thicker padding along the centre of their length than on the margins, that they may make pressure on the space between the shafts of the two bones and keep the fragments parallel. They are fastened by the buckle, loop, or common roller bandage, after which the forearm is supported in a sling.

464. The METACARPAL BONES extend from the wrist to the roots of the fingers. Their fractures are treated on a pasteboard or other splint applied to the palmar aspect and extending along the forearm to the ends of the fingers. It should be well padded, particularly in the hollow of the palm.

465. Small splints on FRACTURED FINGERS may be fastened by strips of rubber plaster.

466. The PELVIS [164] is fractured only by great violence, as in falls from a height to rocky ground or the fal of a horse on its rider. It is often difficult to detect the fracture, the existence of which must be assumed from the nature of the causative violence, the inability of the patient to support himself, a feeling of laceration at the fractured part when any exertion is made, and the absence of any other obvious explanation of these symptoms. The legs of the patient should be bound together to prevent motion, and the utmost care used in transporting him to bed lest displacement be augmented or the pelvic viscera injured by splinters. The catheter should be introduced and retained for some days, to lessen the danger of extravasation of urine in case of possible injury to the bladder, as the escape of urine into the peritoneal cavity would be followed by fatal peritonitis, and into the cellular tissue of the pelvis by diffuse inflammation, sloughing, and abscesses. The patient should be placed in bed in the position in which he is most comfortable, usually on his back, with his knees flexed and supported by pillows under them.

467. Fractures of the THIGH-BONE close to the *hip-joint* occur mostly in aged people, from slight shocks, as in making a misstep. When they occur in younger people they

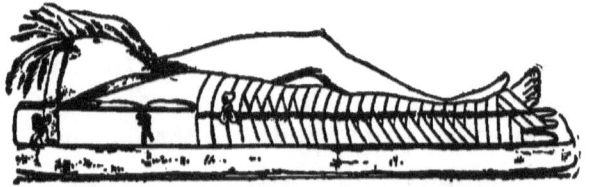

Fractured thigh treated by straight splint.

are usually the result of great violence, and are often complicated with other injuries requiring the most intelligent surgical treatment. In these fractures there is, as in most of the dislocations of the hip-joint, a shortening of the limb, and inability on the part of the patient to move it; but it may be readily lengthened by traction, rotated, or otherwise moved by the surgeon. When the patient lies on his back the foot is usually *everted*, that is, rolled outward on the heel so as to carry the toes away from the central line of the body. Crepitus may be felt by placing one hand over the great trochanter [165] and giving a rotatory motion to the limb with the other. The patient is placed on a hard mattress, and a roller is applied from the toes to above the knee to support the tissues and prevent swelling from the bandage to be subsequently applied higher up. The ordinary wire mattress is too yielding for use in such cases, but when supported beneath by wooden slats it answers the purpose excellently. The splint employed should be long enough to extend from the lower ribs to three or four inches beyond the foot, and it should be provided with two holes near its upper and two notches at its lower end. The limb having been extended, by traction, to its proper length, the splint, well padded, particularly at the ankle, is applied on the outer side. A roller, fixed by several turns around the ankle, is looped around the foot and through the notches at the end of the splint to keep the foot in position; this bandage is then carried upward beyond the knee. Extension and counter-extension are effected by the application of what is called the *perineal band.* A large smoothly folded and well-padded handkerchief is applied to the crotch of the injured side, and the ends carried, one in front, the other behind, the hip to the holes in the splint, through which they are passed and tied firmly together. By this means the splint, and with it the part of the limb below the fracture, is kept down in its proper place. A small wooden splint is applied to the inside of the thigh; and afterward a broad bandage is passed around the pelvis and brought down the thigh to aid in keeping the apparatus in position.

468. Fractures of the SHAFT OF THE FEMUR generally

override each other from half an inch to two inches, owing to the contraction of the powerful muscles connected with the bone. American surgeons usually treat this injury in the straight position by short thigh-splints and *extension by weights.* A roller is applied from the toes to the ankle; a strip of plaster two and a half inches wide is then made to adhere firmly to each side of the limb from the fracture to the ankle, but each strip is cut so long that the lower ends

Fractured thigh in coaptation splints; extension by weights.

may overlap, forming a loop about four inches beyond the sole of the foot; the roller is then carried upward to the level of the fracture. A flat wooden foot-piece, three inches wide and four or five long, is firmly attached within the loop of adhesive plaster, and to it is made fast a cord which passes over a pulley to sustain the extending weights. The weight of the body of the patient affords the counter-extending force, which may be increased by raising the foot of the bed on blocks, or by means of a perineal band extending upward and made fast to the head of the bedstead. Four padded splints are applied to the thigh, and kept in position by loop or buckle bandages; and support is further given to the limb by sand-bags, particularly on the outer aspect, to prevent eversion of the foot.

469. Some surgeons treat fractures of the thigh-bone, particularly of its upper part, in a slightly flexed position, the body being somewhat raised in bed, the thigh and leg supported on a double inclined plane, and the foot, to prevent its eversion, made fast to a suitable foot-piece [471]. Small splints are applied by buckle bandages to the anterior and lateral aspects of the thigh. *Smith's anterior splint* acts the part of a double inclined plane for suspending the

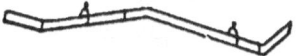

Smith's anterior splint.

limb in the flexed position. It consists of a framework of stout wire, which is applied to the anterior aspect of the limb; short strips of strong muslin or bandage are attached to the wires, and pass beneath the limb, to form a cradle for its support. The splint is then bound on firmly by

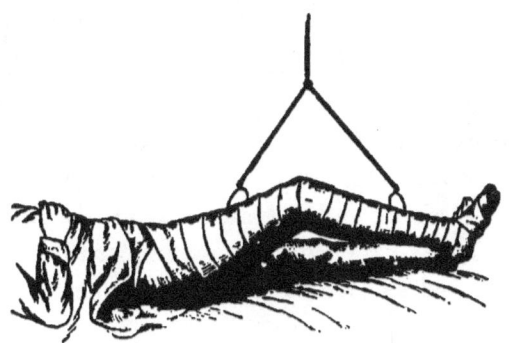

Smith's anterior splint applied.

roller bandages, and the whole suspended as shown in the accompanying illustration. The thigh attachment for suspension should be near the seat of the fracture; the leg attachment about half way between the knee and the ankle.

470. In FRACTURES OF THE LEG, not including gunshot fractures, both bones are usually broken, and in their lower third, their weakest point. Any break in the anterior border of the tibia can readily be detected by the fingers; but

when the fracture is near the ankle-joint the strong muscles of the calf may draw the lower fragments and heel backward, or, if the fracture be oblique, the foot may be displaced outwardly or inwardly. In these cases the absence of dislocation is known by the removal of the deformity under the influence of slight traction. When the fibula alone is broken, as is sometimes the case from a twist of the foot, crepitus may be felt by firmly grasping the parts and moving the broken ends on each other.

471. When, in fracture of either or both bones, there is little deformity, the plaster of Paris, or starch bandage strengthened by a pasteboard splint, may be immediately applied. When the fracture is in the upper third of the leg, the upper fragments are prone to be tilted forward, particularly when the muscles of the anterior aspect of the thigh are put on the stretch by bending the knee. The straight is, therefore, the best attitude for the treatment of such fractures. A roller is applied to the limb, which is then placed on a straight wooden splint, somewhat hollowed on its upper surface, and extending from the middle of the thigh to the heel; after which a pasteboard or gutta-percha splint is moulded to each of the lateral aspects. When the

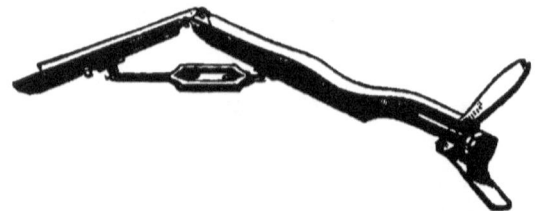

Double inclined plane for fractures of lower extremity.

fracture is below the upper third, the flexed position of the limb is considered to give better results. The double inclined plane for this purpose ends below in a foot-piece, and has a screw beneath the knee-joint by which the angle of elevation may be altered to suit the necessities of the case or to permit of *passive motion* to the joint when the confinement of the limb has continued for a long time. The *fracture-box* is often used for the treatment of these fractures, and is particularly useful in gunshot injuries, and those which are associated with wounds of the soft parts. It consists of a bottom which extends from the knee to the sole, two sides which are hinged so that they

Fracture box for the leg.

may be approximated over the contained limb, and a foot-piece fixed at an appropriate angle. A soft pillow is laid in the box, or oakum, bran, etc., may be used for this purpose; the injured limb is then deposited in it, and the necessary pressure exerted by tying stout tapes or strips of bandage around its yielding sides. If bran is used, a soft muslin cloth should be laid in the box in the first instance, that its ends may be folded securely around the limb and fastened to prevent the leakage of its contents.

CHAPTER XI.

DISLOCATIONS.

472. A dislocation is a displacement of the bones forming a joint; the ligaments are stretched or ruptured, and the articulating surfaces of the bones more or less separated from each other. The bone dislodged is usually that on the lower or distal side of the joint; the other remains in its place.

473. Dislocation is characterized by pain, swelling, loss of power over the part, and deformity from the unnatural position of the dislodged bone. Deformity in fracture may be removed by pulling the parts gently into their proper place, but it returns when the traction ceases; dislocation offers a fixed resistance to efforts of this kind; but once the deformity is removed by the application of a greater force, it does not return. Moreover, deformity in fracture is associated with a mobility in parts which should be rigid, while in dislocation there is rigidity of parts which are naturally mobile. Ordinarily, therefore, there is little difficulty in determining whether an injury is a fracture or dislocation, particularly if the examination is made before inflammatory swelling is developed. *Counter-extension, extension,* and *manipulation* are required to reduce a dislocation. Counter-extension is the force applied to the upper bone to fix it in its place, so that the extensional force applied to the lower bone may be wholly utilized in drawing it away from its unnatural position. Manipulation tilts the lower bone into its place by direct pressure or by some suitable movement of the limb, when the extending and counter-extending forces have freed the bones from their unnatural position. The extending force should be steady, and gradually increased until its object is accomplished. The use of pulleys for multiplying power is seldom required in the recent injuries which constitute surgical emergencies; and chloroform is rarely needful, as the primary shock sufficiently relaxes the muscles.

474. When a dislocation is *complicated* by fracture close to the head of the dislocated bone it is always difficult, and sometimes impossible, to effect its reduction.

475. After a dislocation has been reduced, the joint should be supported by a sling, bandage, or splint for a week or more until the risk of recurrence has been obviated by the recovery of the lacerated or stretched ligaments.

476. The LOWER JAW is dislocated by opening the mouth too widely, as in yawning, etc., or by a downward blow on the chin while the mouth is half open; both sides are usually displaced, sometimes only one. The head of the bone slips from its natural position, immediately in front of the ear, over and in front of the eminence which forms the anterior margin of its articular cavity. The mouth is locked in the half-open position; speech and deglutition are effected with difficulty, and there is a dribbling of saliva from over-secretion by the stimulus of pressure on the salivary glands. To reduce: Two pieces of soft wood, each about as thick as the thumb, are placed, one on each side, between the upper and lower posterior molars, and are held in this position by an assistant standing behind the patient. The operator then places his fingers under the arch of the bone and raises it, gradually increasing the power, by which means the head of the bone is tilted downward and backward over the articular eminence, when the muscles of mastication immediately return it to its cavity with a snap. In the absence of an assistant, the operator may insert his thumbs into the patient's mouth, and press the posterior lower molars downward, while with his fingers he raises

A DISLOCATED LOWER JAW: The condyloid process *a* of the lower jaw has slipped over the articular eminence *b* from its natural position in the articular cavity *c* which lies in front of the canal of the ear. The alveolar arch *d* is thus projected forward and the mouth is open because the coronoid process, *e*, strikes against the lower edge of the malar bone *f*; *g* is the angle of the jaw and *h* its ramus.

the chin; but in this case the thumbs should be guarded by strong leather gloves. A chin bandage [395] should be worn for a few days to support the joint.

477. The most common displacement of the inner end of the CLAVICLE is that in which it is thrown forward from its articulation with the sternum or breast-bone by violence applied to the front of the shoulder. The most common displacement at the outer end of the bone is the removal from it, in a downward direction, of the acromion process of the scapula by direct violence, as by falls or blows on the shoulder. The nature of the injury, in either case, is easily distinguished by the unnatural projection of the end of the collar-bone as it lies beneath the skin. Both luxations are reduced by placing the knee, as a fixed point,

DISLOCATION OF THE SHOULDER-JOINT. 275

between the patient's shoulder-blades and drawing the shoulders steadily backward while an assistant in front of the patient makes pressure to direct the end of the bone into its place. After reduction the retentive apparatus used for fracture of the clavicle [452], with a pad over the displaced end of the bone, should be used to preserve the parts at rest.

478. Dislocations of the SHOULDER-JOINT are more common than those of all the other joints of the body taken together; and what is called the downward displacement of the head of the humerus into the armpit is by far the most frequent of the shoulder luxations; in fact the others are extremely rare as compared with it. The injury is easily recognized: The roundness of the shoulder is lost; its apex is formed by the projecting acromion, beneath which may be felt a depression caused by the absence of the head of the bone from the glenoid cavity of the scapula; the patient supports his forearm, as in fracture, with the hand of the uninjured side; but the arm is rigid, generally in line with the long axis of the body, but with the elbow projecting somewhat from the side. The long axis of the bone is seen to extend from the elbow to the armpit, in which situation its round head may be felt, particularly if at the same time an effort be made to raise the elbow. [In the accompanying illustration the shaft of the humerus is too perpendicular; it should have had a decided outward slope.] Raising or lowering the elbow is difficult and causes great pain, but a slight forward and backward movement

Dislocation downward of the humerus at the shoulder, the head of the bone lying below the glenoid cavity of the scapula.

may sometimes be effected without much inconvenience. This projection of the elbow may indeed be looked upon as a reliable sign of dislocation if it persist when the hand of the patient is laid upon his uninjured shoulder. The fingers tingle and become numb from pressure by the displaced bone on the large nerves traversing the axillary space, and the limb may soon become swollen from interference with the circulation.

479. When recent, this dislocation is easily reduced. An assistant holds the body of the patient fixed; another makes traction on the elbow in the line of the bone, at first downward and outward, and afterward more directly outward by gradually raising the elbow, to draw the humerus out of its abnormal position, while the fingers of the operator in the armpit press outward and upward to help the head over the edge of the glenoid cavity; and as soon as this is accomplished the muscles restore the bone with a snap to its natural position.

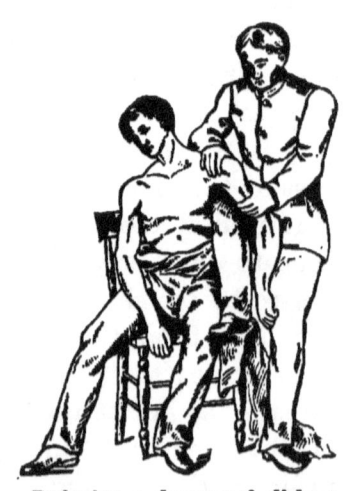

Reducing a downward dislocation of the humerus at the shoulder-joint.

480. Various modes have been suggested of accomplishing reduction when assistance is limited to one person, as, for instance: The patient is seated on a low chair on which the operator places his foot so as to insert his knee into the axilla; with one hand he fixes the shoulder, and with the other grasping the elbow he draws the humerus downward, at the same time endeavoring, by pressing it inward, with his knee as a fulcrum, to tilt its head over the edge of the articular cavity. Again: The patient lies on his back on the ground while the operator, seated by his injured side and facing him, makes traction on the hand and forearm, and with his unbooted foot in the axilla directs the head of the bone to its cavity. Or: The armpit of the patient is hooked over the top bar of a wooden fence and the operator makes traction on the displaced bone, the inertia of the

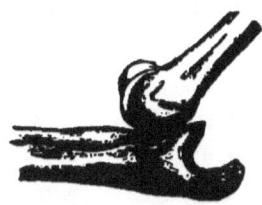

Dislocation backward of the radius and ulna at the elbow.

patient on the other side of the fence affording the counter-extending or resisting force, while the breadth of the top rail, under these conditions, lifts the head of the bone over the edge of the cavity. In muscular subjects it may be needful to clove-hitch [483] a skein of yarn or some other strong soft material to the arm, by which to make the necessary traction. After reduction the arm must be supported in a sling for several days.

481. The most frequent of the dislocations at the ELBOW-JOINT is that in which both bones of the forearm are carried backward and project beyond the lower end of the humerus. The deformity is similar to that in fracture of the lower end of the arm-bone, but it cannot be removed, as in that case, by gentle traction; the forearm and arm are fixed in their abnormal relation to each other, and cannot be unlocked without the intelligent application of suitable force. One assistant fixes the humerus, another makes traction on the

forearm in the direction of its length, and the operator, placing the fingers of one hand in front of the forearm close to the elbow, and those of the other on the projection behind the joint, endeavors to pull the bones away from their position against the lower end of the humerus. Or: The patient being seated on a chair, the operator rests his foot on it and places his knee in the bend of the injured elbow, pressing against the front of the radius and ulna to dislodge them; one hand fixes the humerus while the other makes traction on the forearm, and, after the knee-pressure has been kept up for some time, bends the forearm around the knee to bring the articulating surfaces into contact. The joint and forearm must afterward be supported in a sling.

482. Dislocation of the HAND from the lower end of the bones of the forearm is an extremely rare injury, at one time supposed to be more common because fracture of the lower end of the radius [460] was frequently mistaken for it.

483. The *thumb* has three joints—one opposite the upper part of the ball, one on a level with the web, and a third

Formation of the clove-hitch.

in the free portion. All are liable to luxation, but the middle joint most frequently meets with this injury, the lower bone being usually thrown backward on the upper. To apply a suitable extending force it is sometimes needful to use a skein of thread or a strong tape made fast to the thumb by a clove-hitch, with a few turns of a wetted roller applied beneath to protect the skin. The *clove-hitch* is made by forming two loops as in the cut, and then placing that on the right over or in front of the other, so as to form a double loop; when traction is made on its ends this loop around the thumb does not slip.

484. The joints of the *fingers*, when dislocated, are reduced in the same manner as those of the thumb.

485. DISLOCATIONS OF THE HIP-JOINT are accidents which require the best surgical intelligence for their treatment. In presenting them briefly in this place, it is not suggested that members of the hospital corps should be prepared to undertake their management, but merely that they should have that amount of knowledge of the conditions involved which will enable them to give intelligent assistance to the operator in his efforts at reduction. This knowledge, under the pressure of an emergency which coincides with the absence of better surgical skill, may be of much benefit to the patient.

486. The head of the femur may be thrown out of its cavity in any direction, but systematic writers on surgery divide dislocations at the hip-joint into five classes: 1, Upward on the dorsum or outer aspect of the ilium; 2, upward and backward into a notch between the ilium and ischium; 3, upward and forward toward the pubes; 4, downward and forward below the pubes; and 5, anomalous dislocations which do not properly belong to any of the four classes stated. Sixty per cent of the dislocations of the hip-joint belong to the first class; twenty-five to the second; the remaining fifteen to the other classes.

487. When the head of the bone is thrown upward on the dorsum of the ilium, the hip bulges from the unnatural position of the bone, the limb is shortened, the knee rolled inward and advanced across the lower third of the opposite thigh, and the foot *inverted*, with the great toe resting on the top of the arch of the opposite foot. The patient cannot move the limb, nor can the surgeon straighten it or withdraw it from the other. The head of the bone is directed backward and the trochanter forward. The injury

Dislocation upward on the ilium.

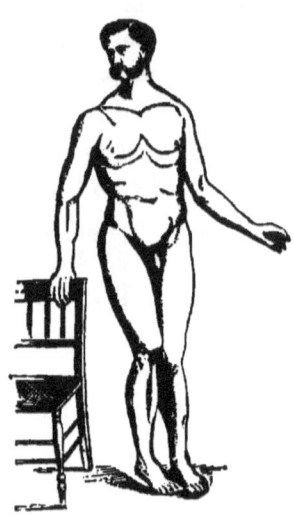

Dislocation upward and backward.

is distinguished from fracture of the neck of the femur by nversion, immobility, and the absence of crepitus.

488. When the head of the bone rests on the ischiatic notch there is less shortening and the knee is less advanced; the limb is rotated inward and the great toe rests on the root of the toe of the opposite foot. The head of the bone is directed backward and the trochanter forward.

489. When the head is thrown upward and forward toward the pubes there is also shortening; but the knee and foot are rotated outward and drawn away from the opposite limb; the roundness of the hip is lost, and the head of the bone, which is directed forward, forms an easily discovered tumor in the groin on the outer side of the femoral artery.

490. The downward dislocation is also characterized by the limb being carried away from that of the opposite side;

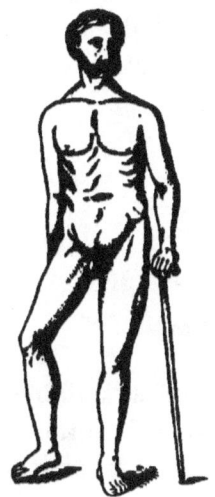 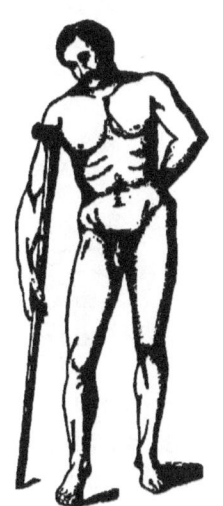

Dislocation upward and forward.   Dislocation downward.

but in this instance there is lengthening; and although the foot is advanced, there is neither inversion nor eversion. The head of the bone is directed inward and the trochanter outward.

491. In dislocation upward the operator places his foot in the perineum as a counter-extending force; he passes the loop of a skein of worsted, which is attached by a clove-hitch to the lower part of the thigh of the patient, over his neck, so that by stretching himself backward he may bring a considerable extending force to bear on the dislocated bone; and when these forces have been continued for some time he endeavors with both hands to rotate the knee outward so as to tilt the head of the bone into its cavity. But generally in these dislocations chloroform has to be administered and steady traction by pulleys used for the purposes of extension. For instance, in the upward dislocation the patient lies on his back; a sheet is folded to form a perineal band, the ends of which are made fast to some immovable object so situated that its resistance shall be exercised in line with the long axis of the femur; a skein of worsted is

Reduction of upward dislocation.

then clove-hitched above the knee, or a suitable leathern band buckled on this part, and hooked on to the system of pulleys, which are so arranged that the extending force may also be in line with the long axis of the femur. When these forces have brought the head of the bone down to its proper level, the operator rotates the knee slightly outward to complete the reduction; or if difficulty be experienced in effecting this, he passes a towel beneath the thigh close to the hip-joint, and tying its ends together, places the bight over his neck and endeavors by raising his body to lift the head of the bone into its cavity while using his hands on the pelvis as a counter-force.

492. In the upward and backward dislocation the patient lies on his side and the extending and counter-extending forces are directed in a line across the middle of the sound thigh.

493. When the head of the bone is thrown toward the

pubes, the patient is placed on his sound side; counter-extension by the perineal band is upward and forward in front of the line of the body, and extension downward and backward behind the line of the body; a towel is used to lift the bone into its place.

494. In the downward dislocation the patient lies on his back; the pelvis is fixed by a sheet which is made fast to

Reduction of upward and backward dislocation.

some immovable object on the sound side; another sheet is passed into the perineum and its ends are carried up between the body and the counter-extending girth, after which they

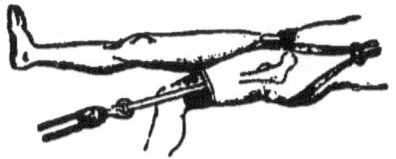

Reduction of upward and forward dislocation.

are attached to the pulleys; extension is made in an upward and outward direction as regards the position of the patient; and when these forces have been sustained for some time, the operator from the sound side passes his hand beneath the patient's sound ankle, and, grasping the ankle of the injured limb, draws it toward him, thereby throwing the head of the bone outward into its place.

495. In all these dislocations a long straight splint [467]

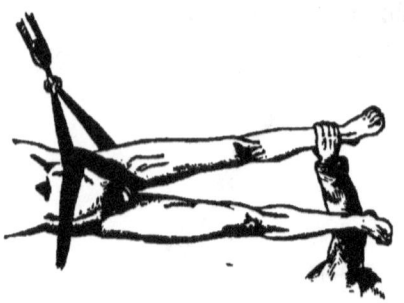

Reduction of the downward dislocation.

must be applied on the outer side, after reduction, and retained for two or three weeks to support the joint and prevent motion.

496. The PATELLA or KNEE-CAP is sometimes driven outward, less frequently inward, from its position in front of the knee-joint. Reduction is effected by bringing the limb forward on the body to relax the muscles on the anterior aspect of the thigh, and then pressing the bone inward or outward, as may be required, to restore it to its place.

497. Dislocations at the KNEE-JOINT are dangerous from the violence needful to cause them and from their inflammatory consequences. The nature of the injury is easily distinguished. Reduction is accomplished by extension from the ankle and counter-extension from the thigh to separate the bones, with pressure by the fingers of the operator outward, inward, forward, or backward, as may be required to bring the head of the tibia into its proper place.

498. The most frequent of the dislocations of the ANKLE-joint is that in which the foot is twisted outward, so that its inner edge rests on the ground. The lower end of the tibia forms a projection on the inner aspect of the joint, and there is a depression on the outer aspect about two inches above the joint where the thinner bone, the fibula, generally gives way. This accident has often been called "Pott's fracture," after the surgeon who first described the true condition of the parts involved. It is treated by extension to the foot, counter-extension to the leg or thigh, and manipulation to return the bones to their natural relationship. The leg is then placed on an inclined plane with a foot-piece, or otherwise supported until the fractured fibula becomes united and the ruptured ligaments healed.

499. Other dislocations at the ankle and of the bones of the FOOT are rare or so complicated by the violence which caused them that the displacement of the bones is only one, and often a secondary one, of the conditions that require surgical treatment.

CHAPTER XII.

THE MANAGEMENT OF CASES OF POISONING.

500. The *strong acids* and *alkalies* can hardly be called poisons. When taken accidentally, they seldom reach the stomach, as their presence in the mouth discovers the accident and leads to the immediate rejection of the mouthful. When taken intentionally, however, the stomach is involved. There is intense thirst with burning and agonizing pain, nausea, and vomiting of matters mixed with darkened blood and portions of mucous membrane. After much suffering the patient becomes exhausted, his pulse rapid and weak, skin cold and clammy, leading to death by collapse. Sometimes the shock of the injury [304] is so great that death may occur with little suffering. In other cases, particularly when ammonia has been used, death may occur from suffocation caused by tumefaction of the larynx. The corrosive agent should be immediately neutralized. Magnesia answers best for the strong acids, as it is bland in itself and disengages no gas to distend the injured stomach; but in its absence chalk, whiting, washing or baking soda may be used, or even soap and water, as the fatty acid of the soap does not interfere with the antidotal power of its alkali. When potash, soda, or ammonia is the corrosive agent, vinegar, lemon juice, aromatic sulphuric acid, etc., suitably diluted, must be administered. Subsequently, the whites of two or three eggs should be given in water, or in their absence, milk, olive oil, or mucilaginous drinks, as gruel, acacia water, or linseed tea. The intense suffering should be alleviated by subcutaneous injections of morphine, and the patient sustained during threatening collapse by injections of brandy, warmth, and friction.

501. Poisons, from the point of view of those who are called upon to rescue the patient, may be divided into three classes.

I. Local irritants which produce an irritation or inflammation of the alimentary canal.

II. Those which combine with the local irritation a certain action on some other part of the system.

III. Those which exert their influence on the nervous system with little or no local action on the alimentary canal.

502. In every case of poisoning the first object of treatment is to rid the stomach of the deleterious agent, and meanwhile, if possible, to prevent its absorption. If free vomiting is occasioned by the poisonous agent itself the use of warm water may suffice to secure its removal; otherwise evacuation is best accomplished by the *stomach-tube* or *stomach-pump*, because the use of these instruments is not attended with the exhaustion that is prone to follow repeated vomiting. To wash out the stomach by siphon action we require the flexible stomach-tube of the stomach-pump case, about a metre of rubber tubing to fit on to the stomach-tube, and a large funnel, holding about two litres, to fit into the other end. The stomach-tube is warmed, oiled, and introduced while the patient is supported in the sitting posture. Its tip, in beginning its descent, should be in contact with the back part of the pharynx. When introduced it is connected with the rubber tubing and funnel, the latter filled with warm water and held below the level of the patient's stomach. When all is ready the funnel is raised that its contents may flow gently into the stomach, and when the latter is distended it may be emptied by lowering the funnel toward the floor. The stomach should be flushed out several times. Should any solid matters block the stomach apertures of the tube they may be cleared out by raising the funnel. The stomach-pump consists of a barrel and piston, the nozzle of the former fitted on its end with a flexible stomach-tube and on its side with another tube for conveying the pumped liquids into a receiving basin; but the valves of the pump are so constructed that the action of the instrument may be easily reversed, thus enabling the operator to pump liquids into as well as out of the stomach. Water should be alternately pumped in and removed until it comes away clear. When the stomach tube or pump are not at hand, *mustard* or *sulphate of zinc* should be preferred as emetics, because they act promptly and with little straining or depression. The dose is one or two teaspoonfuls of the former, or two grams of the latter, in a tumblerful of warm water, vomiting being afterward solicited by tickling the fauces and drinking large quantities of tepid water. *Ipecacuanha*, in two-gram doses, may also be used. Sulphate of copper, from its irritant properties, is likely in many cases to increase the poisonous effect if it be not efficient as an emetic; and tartar emetic induces too much prostration to be used in cases which so often tend to death by the supervention of collapse. *Hydrochlorate of apomorphine*, .006, may be given by *hypodermic injection* when the patient is unable to swallow. In using the hypodermic syringe the exact quantity of the solution to be injected is drawn into the barrel, the needle is adjusted, and the piston pressed lightly until the appearance of the contained liquid at the point of the instrument shows that all air has been expelled. The skin is then pinched up between the forefinger and thumb, and the needle is pushed into the subcutaneous cellular tissue. The injection should be made slowly and gently, and after the withdrawal of the instrument the finger should be kept on the puncture for a few seconds. The arm or forearm is usually selected as the most convenient site, avoiding the vicinity of the subcutaneous veins.

503. The prevention of absorption is effected by the administration of a chemical antidote which will combine with the poison in the stomach to the formation of a substance which is insoluble or inert. This may often be attempted pending the evacuation of the stomach. Tannic acid, for instance, unites with the poisonous alkaloids, and also with tartar emetic, forming insoluble compounds. It is well, therefore, in treating poisoning by vegetable substances, to make the patient drink 250 c.c. of water containing a gram of tannin, or freely of an infusion of green tea, particularly if there is any delay in procuring the stomach-tube or emetic. The tannin is given in the expectation of holding the deleterious agent prisoner until it can be removed; but the necessity for its removal is not obviated, for the tannin compound is soluble and poisonous under certain conditions that might be present in the alimentary canal. So in the case of poisoning by salts or oxids of the metals, arsenic and antimony excepted, it is well to have the patient swallow the whites of two or three eggs diffused in water, for albumen forms insoluble compounds with most of these poisonous substances. If eggs are not at hand, milk or flour and water may be used as less efficient substitutes, as the casein of the one and the gluten of the other have a chemical action somewhat similar to that of albumen. But as these compounds are also liable to be dissolved under certain conditions, those produced with corrosive sublimate and sulphate of copper being, for instance, soluble in an excess of albumen, it is proper to get them out of the stomach as speedily as possible.

504. When the poisonous substance is known, the chemical antidote is sometimes indicated with precision. Thus, when nitrate of silver has been swallowed, a solution of common salt destroys its causticity by precipitating the silver as an insoluble chlorid; when oxalic acid has been used, chalk or whiting neutralizes its irritant qualities by converting it into an insoluble oxalate; carbolic acid and sugar of lead are rendered inert by the use of a soluble sulphate such as Epsom or Glauber's salt; phosphorus is converted into a harmless phosphid by sulphate of copper, and arsenic is deprived of its poisonous properties by the hydrated oxid of iron. In all the cases mentioned except the last, the antidote is administered when available, irrespective of the antecedent or subsequent use of emetics; but in the exceptional case of arsenic, the stomach should be evacuated in the first instance, and the antidote be permitted to remain for the neutralization of any residuum of the poison.

505. Having prevented further injury from the ingested poison by neutralizing and evacuating its unabsorbed remains, the attention of the attendant should be devoted to relieving suffering and *obviating the tendency to death.*

506. The relief of suffering is called for mainly by the irritant class. The burning pain in the stomach and bowels indicates the inflamed or eroded condition of the mucous membrane, and calls for the use of cooling, bland, and protective liquids, as olive oil, mucilage of acacia, linseed tea, barley water, and gruels, with small quantities of ice or ice water to allay thirst. Mustard to the pit of the stomach or large emollient poultices over the abdomen afford relief, but opium is the chief reliance in aggravated cases, given by the mouth, by enema, or hypodermically. As death in these cases is the result of increasing prostration and collapse, the tendency must be counteracted by the use of beef essence, free stimulation, warmth, and frictions.

507. The irritant effects of the few poisonous substances in the second class are controlled as stated in the last paragraph; but oil should not be used as a protective in the case of phosphorus or cantharides, as it dissolves the poisonous principle in both instances, and would probably aggravate the symptoms. The strangury of cantharides is controlled by twenty or thirty drops of liquor potassæ taken hourly in water. Collapse is treated as in the purely irritant class; but in aconite poisoning digitalis may be tried to counteract its directly depressant action on the heart; and if the danger from carbolic acid be rather from failure of the respiration than from collapse, atropine, enemata of coffee, water douches, and artificial respiration should be employed.

508. Of the poisons of the third class, opium, belladonna, and hyoscyamus produce a stupor in which the respiratory wants of the system are unheeded, and the patient dies because he does not respire. Chloral kills in the same way, but there is in addition a paralyzing action on the heart and muscular system that renders strychnine, .003 subcutaneously, of value as a remedy. Prussic acid, hemlock, and tobacco also paralyze the muscular system, and require stimulants, atropine, .001, digitalis, 1 c.c. of the tincture, or strychnine hypodermically to counteract their influence, and artificial respiration to prolong the hope of recovery. Hellebore depresses the heart's action, an effect which is counteracted by digitalis; while the excitement of the heart occasioned by digitalis is quieted by aconite. Strychnine calls for bromid of potassium and chloral, and inhalations

of chloroform or nitrite of amyl to control the muscular spasms and lessen the danger of exhaustion from their continuance, or of death from their intensity.

I. IRRITANTS.

509. ARSENIC causes burning pain in the stomach, with great thirst, nausea, vomiting of brownish matter and blood-stained mucus, tenderness of the abdomen and purging with much griping, straining, and suppression of urine; great prostration with much anxiety, fainting, palpitations, and cramps, ending in collapse with cold and clammy skin, sighing respiration, and imperceptible pulse. The *stomach-tube* or *emetics;* after which a tablespoonful of recently prepared *ferric hydrate*, given every five or ten minutes, until the patient is relieved. The hydrate is prepared by precipitating a solution of tersulphate of iron by liquor of ammonia, and straining and washing the precipitate to free it from excess of ammonia. An antidote, *ferri oxydum hydratum cum magnesia*, the preparation of which involves no loss of time in washing, has been provided by the Pharmacopeia. The tersulphate solution is treated with an excess of magnesia, instead of ammonia, and the precipitate is immediately administered by the tablespoonful, for neither the magnesia nor its sulpahte which is formed during the precipitation interferes with the action of the hydrated oxid. Suffering is then relieved and inflammation allayed by *mucilaginous drinks*, *cataplasms*, and *morphine*, and threatening collapse averted by *stimulants* and the *warmth* of hot-water bottles, flannels, and friction.

510. CORROSIVE SUBLIMATE is similar in its action to arsenic, causing gastric and intestinal inflammation and collapse. *Egg-albumen, milk* or *flour and water;* the *stomach-*

tube or *emetics;* subsequent treatment as in arsenical poisoning.

511. SUGAR OF LEAD induces the usual gastric symptoms, but the vomited matters contain white chlorid of lead; there is great abdominal tenderness with cramps and increasing prostration, but the astringency of the lead salt prevents the purging that generally characterizes irritant poisoning. *Sulphate of soda, sulphate of magnesia,* or *aromatic sulphuric acid* in water; the *stomach-tube* or *emetics;* subsequent treatment as in arsenical cases.

512. NITRATE OF SILVER produces local symptoms with much constitutional disturbance. *Common salt* in solution; the *stomach-tube* or *emetics,* and subsequent treatment as in arsenic.

513. ZINC SALTS.—The sulphate, taken sometimes by mistake for Epsom salt, is its own antidote by *emetic* action; subsequent collapse requires treatment as in arsenic.

514. COPPER SALTS.—The primary effects are those of local irritants; the sulphate may be its own antidote. *White of egg, milk,* or *flour and water;* the *stomach-tube* or *emetics* if required; subsequent treatment as in arsenical poisoning.

515. TARTAR EMETIC.—Collapse is hastened by a direct depressant action on the circulation. This poison acts as its own antidote in evacuating the stomach. *Tannin* in solution, or *infusion of green tea,* with *morphine* to allay excessive vomiting and relieve suffering, and *stimulants* as required.

516. OXALIC ACID.—*Chalk* or *whiting* diffused in water; the *stomach-tube* or *emetics,* and subsequently measures to relieve suffering and avert, or recover from, collapse.

517. CROTON OIL, COLOCYNTH, and other DRASTIC CATHARTICS tend to death by exhausting the patient. The *stomach-tube* or *emetics; opium* and *stimulants.*

518. COLCHICUM causes violent local irritation with intense prostration. *Tannin* or *green tea;* the *stomach-tube* or *emetics; morphine* and *stimulants.*

II. IRRITANTS WITH A SPECIFIC ACTION.

519. CANTHARIDES produce the gastro-enteric symptoms of irritant poisoning; and also an irritant action on the kidneys, manifested by strangury and bloody urine; death is preceded by delirium and convulsions. The *stomach-tube* or *emetics; mucilaginous,* but not oily, preparations; *morphine* and *stimulants.*

520. ACONITE causes a burning feeling in the throat and stomach, with nausea, vomiting, purging, and intestinal pains; numbness and tingling of the muscles, merging into paralysis; great depression of the circulation and prostration, ending in general collapse, sometimes with convulsions and stupor. The *stomach-tube* or *emetics; stimulants* by the rectum, or subcutaneously if rejected by the stomach; *warmth* and *frictions; tincture of digitalis,* 1 c.c., hypodermically, and repeated, if need be, to strengthen the heart's action.

521. PHOSPHORUS manifests its irritant properties by heat and swelling of the tongue and throat; pain and distention of the abdomen, with mucous or bilious vomiting and purging, sometimes bloody; anxiety, restlessness, cramps, and convulsions, ending in stupor or general collapse. *Sulphate of copper,* .2 in a tumbler of water every five minutes until vomiting is induced; if free vomiting have already occurred, the copper should be given once, with opium to promote its retention; *sulphate of magnesia* to carry off the poison by its cathartic action; *mucilaginous,* but not oily, drinks; *cataplasms, opiates,* and *stimulants* as required.

MANAGEMENT OF CASES OF POISONING. 295

522. CARBOLIC ACID occasions a burning feeling, whitening the mucous membrane of the throat and stomach, and speedily rendering it insensitive by coagulating its albuminoid principle; intense depression, cold and clammy skin, feeble pulse, insensibility quickly deepening into stupor, with death from failure of the respiration or of the heart. *Sulphate of soda* or *sulphate of magnesia, white of egg, milk or flour and water; stomach-tube* or *apomorphine subcutaneously*, as emetics when introduced into the stomach have little influence on its disorganized mucous membrane. *Stimulants* by enema, or *ammonia* or *brandy* by hypodermic injection, with *warmth* and *frictions* if syncope threaten; *atropine* and *artificial respiration* if the breathing fail.

III. POISONS ACTING ON THE NERVOUS SYSTEM.

523. OPIUM NARCOTISM. — When a poisonous dose of opium has been taken, the patient speedily passes from drowsiness into a deepening stupor, in which his nervous centres ultimately fail to recognize the necessity for respiring, and he passes into death by ceasing to breathe. At first the face is flushed and the breathing stertorous, the pulse full and slow, and the stupor profound; but this condition differs in several points from apoplexy. There is no paralysis other than the temporary relaxation and loss of power involved in the loss of consciousness; the pulse, although full, is soft and easily compressed, the pupils are contracted, and the patient may be aroused to some degree of confused consciousness by shaking, slapping, douching with cold water, and speaking to him in loud, sharp tones. After a time the pulse becomes weaker, small, scarcely perceptible; the respiration diminishes in frequency by long intervening pauses, and inspiration is slow and often interrupted; the face becomes pale, the skin clammy and cool,

the extremities cold. In the early stages the patient should be douched with cold water, to rouse him from his stupor; the stomach-tube should be used or an emetic of mustard or sulphate of zinc administered, after which strong coffee is swallowed from time to time, or administered by enema, to counteract the narcotic tendency. Every effort must be made to keep the patient awake and conscious of the necessity of breathing. If the stomach-tube is not at hand and an emetic cannot be taken by the mouth, a hypodermic injection [502] of .006 apomorphine may be used. The respirations should be closely watched, and if they are found to be steadily decreasing, and to have fallen below ten per minute, sulphate of atropine .001 should be given by subcutaneous injection, and repeated once or twice at intervals of ten minutes if no favorable result be observed in the mean time; but if the respiration becomes more frequent, or even more regular, and particularly if the contracted pupils undergo a slight relaxation, the patient may be permitted to rest without further effort on his behalf, unless called for by a subsequent impairment of the regularity or frequency of the breathing. Permanganate of potassium, .4, is recommended by some as a valuable antidote; the water used in washing out the stomach might have this quantity of the permanganate dissolved in it. Artificial respiration may be used as a last resource.

524. Chloral.—The chloral sleep deepens into stupor, during which respiration fails and the pulse becomes weak and small. Treatment as in opium poisoning, with sulphate of strychnine, .003, by hypodermic injection, if death from heart failure seems impending.

525. Belladonna and Hyoscyamus.—Dryness and constriction of the throat, burning in the stomach, dimness of vision, with dilated pupils, hurried breathing, headache, and delirium ending in stupor, with feeble pulse, cold extremities, and diminished respiration; sometimes a scarlet rash on the skin. *Tannin;* the *stomach-tube* or *emetics; coffee; mustard* and *friction* to the extremities.

526. PRUSSIC ACID, CYANIDE OF POTASSIUM. — Death occurs quickly from paralysis of the respiration and heart. *Cold douche* and *ammonia* to the nostrils; *artificial respiration* and *frictions; enemata* of *brandy.* When OIL OF BITTER ALMONDS is the poisonous agent, there will probably be time for the use of *emetics*, with subsequent treatment by *stimulants.*

527. HEMLOCK.—Extreme muscular prostration and death from paralysis of the respiration; no delirium or coma. *Tannin* or *green tea;* the *stomach-tube* or *emetics; stimulants; frictions; atropine* and *artificial respiration.*

528. TOBACCO induces dizziness, confusion of ideas, faintness, nausea and vomiting, intense muscular prostration, and fatal collapse with or without stupor. *Tannin; stomach-tube; stimulants, warmth, frictions,* and *strychnine* by subcutaneous injection.

529. HELLEBORE. — Intense prostration, imperceptible pulse, cold, clammy skin, nausea and attempts at vomiting, fainting, and fatal collapse. *Tannin* or *green tea; stomach-tube* or *emetics; morphine* and *stimulants* by the mouth or rectum; *ammonia, frictions,* and *digitalis.*

530. DIGITALIS.—Nausea, vomiting, and great prostration, with cold sweats and feeble, almost imperceptible pulse, stupor or delirium, and death from spasm of the heart. *Tannin* or *green tea;* the *stomach-tube* or *emetics; stimulants* and one or two minims of *tincture of aconite* by hypodermic injection.

531. STRYCHNINE.—Violent spasms of the muscular system, lasting from one to five minutes, with intervals of

relaxation, prolonged sometimes for half an hour. During a convulsion the features are contorted, the body bent rigidly backward, and the breathing stopped or much impeded; brain not affected. The patient dies of asphyxia during a convulsion, or exhausted by the frequency and violence of the fits. *Tannin; stomach-tube* or *emetics; bromid of potassium*, 7 gm., and *chloral*, .7, to control the paroxysms, with *chloroform* or *nitrite of amyl* inhalations when the convulsions are specially severe; *artificial respiration* as a last resource.

532. POISONING BY ARTICLES OF FOOD.—Meat, sausage, cheese, shell-fish, certain species of mushrooms, fruits, etc., sometimes occasion nausea, vomiting, violent cramps, pain, diarrhea, and great depression of the vital powers. The *stomach-tube* or *emetics* and *cathartics* to remove the noxious material, with *stimulants* to relieve the subsequent depression.

533. POISONING BY RHUS TOXICODENDRON.—A cutaneous inflammation somewhat resembling erysipelas [372] is often occasioned by contact with or even by exposure to emanations from the leaves of the *poison oak, poison ivy,* and other species of Rhus. The poison oak is a small shrubby plant with a leaf which consists of three leaflets: two lateral, each about four inches long and two-thirds as broad, springing directly from the leaf-stalk, ovate and pointed at the apex, and one terminal, stalked, ovate, with a wedge-shaped base and pointed apex—all notched or lobed on the margins and downy on the under surface. The poison ivy climbs on trees and rocks to a considerable height; its leaflets are usually smooth above and below and entire on the margins. The head and face become affected, and sometimes the hands and upper extremities, or the hips, scrotum, and adjoining parts of the thighs. The affected

surface is of a lurid-red color, covered, in aggravated cases, with vesicles, the contents of which become incrusted on drying, or ooze as a thin liquid from superficial fissures. Pain, heat, irritation, and swelling are associated with the redness; but although these often occasion considerable suffering and loss of sleep, there is seldom much constitutional disturbance. Generally the inflammation subsides in about a week without leaving subcutaneous suppurations. It is treated with aperient medicines and cooling lotions.

CHAPTER XIII.

DISINFECTANTS AND THE MANAGEMENT OF INFECTIOUS DISEASES.

534. DISINFECTANTS are to be used only for a specific purpose; in the absence of any infectious disease they are not required, and their expenditure for purposes of general post sanitation is not authorized. Sulphate of iron and other cheap deodorants and antiseptics may be used when necessary, but the necessity for their use is regarded as a reproach upon the sanitary police of a post.

535. *Boiling in water* may be relied upon to purify articles of body-clothing and bed-linen which have been exposed to the infection of disease. All infected articles must of course be kept separate from the general wash.

536. Solution of *corrosive sublimate,* one drachm to the gallon of water, *chlorinated lime,* one ounce to the gallon, solution of *chlorinated soda* diluted with nine volumes of water, *carbolic acid* in a two-per-cent. solution, and trikresol, also in a two-per-cent. solution, are efficient disinfectants in certain cases. They may be used for washing the floors, woodwork, walls, etc., of rooms; they may be used also for the hands, and, with the exception of the sublimate solution, for general personal use and for disinfecting soiled linen before transferring it to the laundry. Stronger solutions should be employed for the treatment of excreta or for saturating the sheets in which a dead body is to be enveloped pending the arrangements for burial. A suitable strength is obtained by dissolving two drachms of corrosive sublimate or four ounces of chlorinated lime in a gallon of water, or by making a five-per-cent. solution of carbolic acid. Any of these should be added, in volume equal to that of the material to be disinfected, to each dejection in cases of typhoid fever, cholera, yellow fever, and epidemic dysentery, to the matters vomited in cholera, yellow fever, scarlet fever, and diphtheria, and to the expectorations in the two last-named diseases and typhoid pneumonia. The disinfectant should be permitted to act on the excreta for an hour, after which the contents of the vessel may be disposed of, preferably by burial. *Quicklime* prepared as milk of lime, 1 part to 8 or 10 of water, may be used for disinfecting excreta. As contact with metals decomposes the sublimate solution, it should be used only in wooden, earthenware, or other suitable vessels. *Formalin* has lately been used as a disinfectant, in solution and also in the form of gas, but it has not been placed on the Army Supply-Table.

537. Privy-vaults or cesspools that have become infected by the addition of discharges direct from the patient must be treated liberally with milk of lime, chlorinated lime, or strong solutions of sublimate or carbolic acid.

538. A *dry or oven heat* of 230° Fahr., continued for two hours, is useful in dealing with infected garments which would be injured by immersion in boiling water or disinfecting solutions.

539. Free exposure to *flowing steam* for an hour or more is an efficient disinfectant for clothing, and the only one that can be satisfactorily applied to mattresses. Clothing should be unfolded and mattresses uncovered and freely opened up to penetration by the steam.

540. *Fumigation with sulphur* has been extensively used by municipal health officers in the disinfection of rooms, particularly after small-pox, and by quarantine officers in purifying yellow-fever ships. This fumigation is directed to the destruction of infection in the crevices of the floors, walls, etc., all of which should be carefully cleared of dust before their exposure to the fumes. The dust thus collected should be burned. Many articles may be conveniently disinfected without injury to their texture or color during the fumigation of the room in which the patient was treated. Each article should be unfolded and freely exposed to the sulphurous acid. The fumigation should last for twenty-four hours, during which three pounds of sulphur should be consumed for every thousand feet of cubic space. Articles of metal should be removed or covered with grease to protect them from the fumes. Due precaution must be taken against danger from fire, as by placing the sulphur in iron vessels bedded in sand; and the room must be made as air-tight as practicable by closing all the chinks.

541. *Destruction by fire* should be resorted to only when disinfection by other means *would cost more than the value of the articles*, as in the case of soiled dressings, clothes, and bedding that are so worn out as not to be worth the trouble of disinfecting, mattresses and pillows that would require the penetration of superheated steam to give reliable results, and tents when the contagious disease that called for their use is at an end.

542. When a contagious disease appears in a command, it is important that it should be recognized at the earliest possible moment.

543. ITCH OR SCABIES usually makes its appearance on the hands, in or about the clefts of the fingers, and from this it spreads over the body and limbs, affecting chiefly the flexures of the joints and other parts where the skin is thin; the head and face are seldom involved. It consists of distinct reddish points with a minute bead of liquid at the apex; but as the intolerable itching leads to scratching and subsequent inflammation, its vesicular character is often obscured. It is caused by a minute insect, the *acarus scabiei*, which burrows in the skin near the vesicles. When the insect is killed the inflammation subsides immediately. Sulphur ointment at night, with a thorough scrubbing in the bath-tub in the morning, repeated on two or three successive days, effects a cure. Clothing, bed-linen, and blankets should be boiled to destroy their contagion.

544. A disease called ARMY or CAMP ITCH must be distinguished from the parasitic scabies or itch of civil life. It is developed in hot weather during an active campaign when neither time nor facilities for personal cleanliness are available. The over-stimulation of the skin during the perspiration of hurried marches, combined with the rough contact of coarse-fibred underclothing, gives rise in some sensitive constitutions to an eruption of minute papules which itch intolerably, particularly during the night, when the attention of the individual becomes fixed on his cutaneous sensations. This condition is known as *prickly heat*. In aggravated cases the papules are torn during the efforts of the patient to find relief by scratching, and the surface presents many dried points of dark blood. Dust and dirt accumulate on the irritated skin and intensify the irritation. The *pediculus corporis* or body louse may appear under conditions of filth and overcrowding, and the irritation of its presence augments the mental and bodily disquietude of the affected individual. When any or all of these irritative causes produce in some parts papules with darkened summits, and in others scabs and crusts of dried, purulent matter with a thin, acrid liquid oozing from their cracks and fissures, the disease is called camp itch. It affects the chest, abdomen, and outer aspect of the limbs, while scabies is generally found in the clefts of the fingers, the flexures of the joints, the inner aspect of the limbs, and other similarly protected parts. Rest in hospital with baths, the removal of all sources of irritation, and lead lotion or some antiseptic ointment, will usually control the disease.

545. Three species of *pediculi* occasionally find favorable conditions for their development on the uncared-for human surface. One of them, that mentioned in the preceding paragraph, will sometimes in a prolonged campaign appear in a squad or company, spreading from man to man by the contact of blankets or baggage. Its body is whitish, elongated, and somewhat flattened. It makes its abode and deposits its ova or nits along the seams or folds of the inner garments. The other species are rarely seen in military commands. They are the *pediculus capitis*, which infests the scalp, and the *pediculus pubis*, which may spread from its nominal locality over the whole of the surface except the scalp. The one is smaller in size than the pediculus corporis, but is otherwise of similar appearance; the other is square-shaped, flattened, and provided with crab-like claws. Both deposit their nits about the roots and stems of the hair. Personal cleanliness with the application of red precipitate or mercurial ointment, and the boiling or burning of infected clothes, will speedily free the individual from such undesirable companionship.

546. SMALL-POX.—When an individual becomes affected with headache and fever two weeks after he has been exposed to the contagion of small-pox, his case should be regarded as suspicious, particularly if there be much pain in the loins and obstinate vomiting. On the third day of the fever small reddish points appear on the forehead and nose. Next day similar points are found on the neck, chest, and arms, while those on the forehead have become darker in color and larger, feeling like small shot under the skin; meanwhile, as the eruption comes out the fever abates. These developments authorize the immediate isolation of the patient under the care of special nurses, who should be protected by fresh vaccine lymph, even though they bear the scars of small-pox itself. The command should then be inspected with the view of protecting by revaccination those who have not recently undergone the operation. By *isolation* is meant the transfer of the patient to a separate ward, room, or tent, where he will have communication with none save those in attendance, and where the intercourse of the latter with the outside shall be so regulated as to prevent the transmission of infection.

547. *Vaccination* is effected by punctures, scratches, or abrasions. When lymph is taken from the vesicles of a healthy vaccinated child on the eighth day of its vaccination the operation is said to be *arm-to-arm* vaccination. This is the best way of dealing with children under ordinary circumstances. But when large numbers of persons have to be vaccinated without delay on account of probable exposure to contagion, fresh lymph must be obtained in quantity, dried on quills or ivory points in the form of crusts, or made up into cones. Vaccine lymph derived from a vesicle on the arm of a vaccinated person is said to be *humanized;* when obtained, as it generally is, by artificial cultivation on the calf, it is called *bovine* lymph. In operating by puncture, which is suitable only for persons with thick skins, a somewhat blunt lancet is run horizontally under the cuticle or scarf skin for about 2 mm. Three such punctures are made close to each other, and three others at a distance of 3 cm. from the first set. A little of the lymph is then inserted into each puncture. If ivory points are used, the lymph must be moistened with water before insertion; fragments of crusts or cones must be rubbed up with a drop or two of water to the consistence of cream. Instead of punctures, two sets of light scratches may be made, each covering a space about 5 mm. in diameter and 3 cm. apart, into which the lymph is rubbed with the flat of the lancet, ivory point, or quill. In the case of tender skins which bleed readily, it is better to scrape the surface at the two points until the cuticle is removed, and then rub in the lymph as into the scratches. The outer aspect of the left arm is usually selected as the site for vaccination. In exposing this site it is advisable as a rule to drop the clothes from the shoulder rather than to roll up the sleeves, for the latter if tight will by their pressure engorge the arm, promote the oozing of blood, and prevent absorption.

548. On the third day after the insertion of the lymph in a primary or first vaccination, a slight reddish tumefaction is observable, which next day is tipped with a little clear lymph. This vesicle increases in size until the eighth day, when it becomes *umbilicated* or depressed in the centre and surrounded by an areola of cutaneous inflammation. The lymph is now mature and fit for use in other cases. By the tenth day the inflammation extends an inch or more in all directions from the vesicle, which loses its pearly color and becomes turbid and purulent. At this time there is usually a slight feverishness, and the glands in the armpit may be swollen. In a day or two the pustule breaks and dries up into a brownish crust, the inflamed areola meanwhile disappearing. The crust falls off about the twenty-first day, leaving a characteristic, slightly depressed cicatrix.

549. When one who shows scars of a successful vaccination is subjected to revaccination, the progress of the vesicle as above described is often modified by the influence of the primary vaccination. The vesicle and its areola may be smaller and the crust fall off at an earlier date. Vaccination protects from small-pox; but as its influence fades in time, it does not continue to protect completely, although it renders the disease, which is then usually called *varioloid*, less protracted and less dangerous to the individual than it otherwise would have been. So vaccination protects from revaccination; but as its influence fades, it does not protect completely, the resulting vesicles being modified as small-pox would have been modified had the individual been infected with the matter of the small-pox pustule instead of with that of the vaccine vesicle. These modified results of revaccination must therefore be recorded as · successful operations because they exhaust the existing susceptibility to small-pox.

550. The convalescent from small-pox should be kept in isolation until the skin is free from crusts. Infected clothing or bedding should be destroyed by fire or disinfected by boiling, steam, or sulphur fumes. Tents, when not required for further use as pest hospitals, should be burned. Infected rooms are disinfected by fumigating with sulphur or by cleaning away all dust, washing with solution of corrosive sublimate or chlorid of lime, and freely ventilating.

551. CHICKEN-POX.—It is sometimes of importance to distinguish between chicken-pox and small-pox or so-called varioloid. The onset of both is by fever; but in small-pox the fever is severe and lasts for three days before the *papules* appear, while in chicken-pox it is mild and precedes the eruption of *vesicles* by twenty-four hours only. The vesicles of small-pox are umbilicated and take nine days to mature; those of chicken-pox are usually rounded and begin to dry up by the end of the third day. Small-pox lasts

three or four weeks; chicken-pox runs its course in eight or nine days.

552. The chicken-pox patient should be isolated from other children until the skin is free from crusts or scabs. The clothes and bed-linen should be disinfected by boiling; the room by thorough cleaning, swabbing with solution of corrosive sublimate, and free ventilation.

553. SCARLET FEVER manifests itself from three to five days after exposure to the contagion. Marked feverishness, redness of the throat, and perhaps pain or difficulty in swallowing are its first symptoms. At the end of twenty-four hours an eruption of small scarlet dots appears on the neck and chest, and afterward on other parts of the surface, coalescing first into large, irregularly shaped patches, and by the end of the second day into a generally diffused efflorescence. The eruption lasts from four to six days, during which the fever continues and the throat affection becomes aggravated, the fauces swollen, and the tonsils covered with soft, yellowish sloughs.

554. A scarlet-fever patient should be carefully isolated, no matter how mild the disease may be in that particular case; for although mild cases are sometimes called *scarlatina*, their contagion is as dangerous to others as that of the most aggravated case of the disease. Isolation should be kept up until the desquamation of the skin has been completed. Bed-linen and body-clothing should be boiled, and the room and all articles that would be injured by boiling fumigated with sulphur, the walls and floors being subsequently treated with corrosive sublimate or chlorid of lime in solution.

555. ROSEOLA.—It is sometimes impossible to distinguish cases of this unimportant rash from mild cases of scarlet ever. The appearance of the eruption is similar in both, and in both there may be no other symptom on which to base a diagnosis. In uncertain cases it is best to isolate as if scarlet fever were under treatment [557].

556. MEASLES.—About eight days after exposure to the contagion, the patient becomes affected with feverishness, cold in the head, and sometimes also in the chest. When these have continued for three or four days, the eruption appears on the face, and in the course of four days more it has spread over the neck, chest, abdomen, and limbs; but by the time the later crops are appearing the earlier are already fading. The eruption consists of small dots, which coalesce into irregular-shaped patches of a dull red color, associated with slight tumefaction, particularly on the face. The patient should be isolated until the fine scales of desquamation have been completely shed. Clothing and bedding should be disinfected by boiling, and rooms by thorough cleaning, washing with sublimate or fumigating with sulphur.

557. ROTHELN or GERMAN MEASLES bears the same relation to measles that roseola bears to scarlet fever, or chicken-pox to small-pox. When these mild diseases occur epidemically, there is no difficulty in showing their differences from the dangerous eruptive fevers which they resemble; but when the first cases occur, the doubt as to their character often calls for careful isolation and treatment until their true nature has been revealed by further observation.

558. TYPHOID FEVER [79, 80].—The infection of this fever resides in the stools of the fever patient. Laundresses frequently contract the disease from contact with soiled linen. The discharges of a typhoid patient should be received into vessels containing corrosive sublimate or chlorid of lime; soiled bed-linen and clothing should be steeped in a solution of chlorinated lime or soda before being removed from the ward; the floor and furniture, when tainted in any way, should be immediately washed with one of the solutions mentioned, and the person of the patient sponged when necessary with a dilution of chlorinated soda. The ward, as in all infectious diseases, should be freely ventilated. As the disease is propagated chiefly by the introduction of its infection with food or drink, care should be exercised by attendants and others in purifying the hands after contact with infected articles, particularly before eating, and obviously no food or drink should be used in the room or ward save by the typhoid patients. When the disease prevails as an epidemic, its spread is usually occasioned by an infected water supply. In this case water must be obtained from other sources, and be carefully guarded from infection; but if a natural pure supply be not available, the water must be boiled, distilled, or filtered efficiently before being used as a drink.

559. CHOLERA AND YELLOW FEVER [82].—All the measures suggested for limiting the spread of typhoid fever are applicable in the management of cholera cases. In the presence of an epidemic, the purity of food and drink requires every care. With the exception of fruits that are protected by an outer rind, no article of food should be eaten that has not been subjected to heat in its preparation for the table. Disinfectants should be freely used in the sinks, as the dejections of choleraic diarrhea are as infectious as the rice-water discharges of defined cholera. It is well, also, to use similar disinfectant and protective measures when dealing with yellow fever.

560. As in both these diseases the locality appears to become infected, security can sometimes be attained only by promptly withdrawing from the dangerous region. If troops fall back before the advance of the disease, or pass

to one side of the line of its advance, they will escape seizure, provided judicious quarantine restrictions are imposed on intercourse with infected localities. When, however, such a retreat is incompatible with strategic requirements, a line of sentinels should be posted around the camp to cut off all unauthorized communication. No person should be admitted within the lines without undergoing quarantine examination; no baggage or supplies without being disinfected or passed by the quarantine officer. Water-supply, if it comes from without, or if a suspicion of contamination from without can be harbored against it, must be boiled before being used, even for police purposes; while any pure supply should be placed under guard for use as drinking water. Should it be impossible to procure such supply by other means, it ought to be boiled, distilled, or filtered through germ-proof filters. Police regulations should be scrupulously carried out. The troops should be protected from all antihygienic influences. There should be no unnecessary exposure to sun, rain, or night air, and no drills or fatigue duties other than to furnish occupation and needful exercise. When the military conditions permit, the camp should be viewed as engaged in an active campaign against an insidious and implacable enemy, and the attention of every officer devoted to superintending the conduct of his men with special reference to this view.

CHAPTER XIV.

PHARMACY.

561. All the members of the hospital corps are instructed in the elements of pharmacy. Those who show a bent for this kind of work are afterward assigned to duty in the dispensary. The course of instruction teaches the care of the dispensary, the names and uses of the utensils and apparatus, and the methods of cleaning them; the care of the scales, and the use of the metric system of weights and measures. The student is then required to familiarize himself with the various articles on the shelves of the dispensary and the pharmaceutical processes connected with them, as solution, evaporation, precipitation, filtration, fusion, sublimation, distillation, etc.; the making of pills, mixtures, tinctures, ointments, etc.; and the use of the thermometer, hydrometer, urinometer, etc. Poisons and their antidotes [p. 286], and disinfectants, with their uses in limiting the spread of infectious diseases [p. 300], also form part of this course.

CHAPTER XV.

ELEMENTS OF COOKERY.

562. Lectures are given to hospital corps companies of instruction on the articles of the ration, their money value in connection with the accumulation of a hospital fund, the principles involved in cooking them and in arranging for variety in bills of fare; but a knowledge of practical work in the preparation of food can be obtained only by experience in the kitchen. The "Manual for Army Cooks," published under the direction of the Commissary General of Subsistence,[1] gives full instructions on the subject.

563. All substances used as food are of an organic, and hence putrescible, nature. If bacterial germs are kept away the substance will not putrefy, but will dry up like the preserved meat strips in use in Mexico and other hot countries. The germs of putrefaction are so constantly present in the atmosphere that it is difficult to preserve food for any time when heat and moisture also are present; but when there are unclean spots in a kitchen or store-room the speedy putrefaction of exposed articles becomes a certainty. Cleanliness is therefore the first requisite in a cook; no excuse should hold good against a want of cleanliness.

564. The application of heat in the process of cooking promotes the breaking up of the organic tissues and facilitates their digestion. Moreover, as heat destroys all germs,

[1] Washington. Government Printing-office, 1896.

the thorough cooking of food is preventive of intestinal worms, trichiniasis, etc., and lessens the danger of seizure in times of epidemic disease. The albuminoids, when introduced into the stomach in their raw state, undergo coagulation with subsequent solution of the coagulum in the gastric juices. Heat applied in cooking coagulates the albuminoids and prepares them for immediate solution. At the same time it disintegrates or gelatinizes the fibres and permits the dissolving process to penetrate more readily into their substance. Starch, which forms so large a proportion of all vegetable food, is indigestible in its raw state, but under the influence of heat its granules are broken up and its solubility increased by its transformation into dextrin and glucose.

565. Heat is applied in cooking by various methods, as boiling, stewing, frying, broiling, roasting, and baking.

566. The water used in *boiling* is merely the medium by which the heat is communicated. None of the juices of the meat or other article should be permitted to waste into it. To this end the water should be slightly salted and boiling vigorously when the meat to be boiled is dropped into it. After five minutes of active boiling the process is conducted at a slow boiling or *simmering* temperature for a period varying with the weight of the meat, usually estimated at about fifteen minutes per pound. Salt meats, however, should be put on the fire with cold water. Over-boiled meat, like a hard-boiled egg, is tough from the density of its coagulation. Fish is boiled in like manner until the meat is found to separate readily from the back-bone, after which it should be removed from the water and drained, else it will become water-logged and insipid. Potatoes and other root vegetables should be cut into pieces of equal size, that the heat may penetrate to the centre of each at the same time. If potatoes are peeled before boiling, the water should be strongly salted to prevent the antiscorbutic salts from being dissolved out. When a fork passes easily to the centre, the water should be poured off and the vessel replaced on the fire to get rid of superfluous moisture by steaming. Beans, peas, rice, and other hard grains require soaking in water before boiling. Milk, particularly if thickened by the swelling starch cells of any of the meals or flours, should be stirred constantly to prevent the scorching of the bottom layer; preferably, such preparations should be made in a vessel having a water jacket.

567. In *stewing* meat, the small quantity of water used is intended to be a part of the food. The meat is chopped up to facilitate its disintegration, and the boiling is conducted slowly and with care, to avoid scorching.

568. In making *soup* or *beef stock*, it is intended that the whole of the soluble organic substances shall be transferred to the water. Gelatin, dissolved from bones by seven or eight hours' continued boiling, forms the basis of the soup; but this is usually strengthened by the addition of meat which, from its toughness, cannot be disposed of otherwise to advantage. Gelatin is held by some physicians to be of no value as a nutriment, but this seems inconsistent with the results of every-day experience. Simmering, rather than active boiling, is required in making soup; froth and grease should be skimmed off as they gather on the surface. The stock should be kept for use in tin or earthenware vessels; iron vessels give it an unpleasant taste. The soup is completed by adding vegetables, cut into small dice, to flavor and give substance.

569. In making *beef tea*, lean meat is chopped up finely and simmered with water, a pint to a pound. *Beef essence* consists of the juices collected by digesting the chopped meat in a water bath. The meat is usually placed in a wide-mouthed bottle, which is securely corked, and cooked in a kettle of boiling water.

570. Hot fat or lard is used as the means of communicating heat in the process of *frying*. The melted fat should be hot enough to give out occasional puffs or jets of smoke before it is ready for the immersion of the articles to be fried. The transmitted heat cooks the interior, and none of the fat penetrates unless the temperature be too low. Meats should be rolled in dry crumbs, cracker dust, flour, or corn-meal. Food is also said to be fried when it is cooked in a pan with merely enough of fat to prevent it from becoming scorched and sticking to the hot metal. A fork should not be used in turning meats, as the holes made by it permit escape of the juices.

571. *Broiling* is generally conducted over a clear, smokeless fire by means of a supporting gridiron; but if it be done *before* instead of *over* the fire, the juices of the meat may be collected in a dish and used for dressing.

572. *Roasting* is properly conducted in front of a clear, hot fire, the roast swinging round by a mechanical contrivance to present every part in sequence to the radiated heat. A polished concave reflector on the off side concentrates the rays on the meat, and a dish placed underneath collects the liquid drips, which should be spooned up from time to time and flushed over the roast.

573. *Baked* meats are now generally called roasts. The oven is at a proper heat when the arm, bared to the elbow, can be thrust in and held for about fifteen seconds without discomfort. The meat, properly skewered and placed in a roomy pan, requires frequent attention while in the oven. If it be not basted or flushed with the melted fat from the pan at short intervals, the juices from the interior will be evaporated, and the roast will become dry and insipid. Meats are baked about fifteen minutes to the pound.

574. If flour be mixed with a little water into a stiff, plastic mass, and then kneaded and teased with the fingers, dipping it into water from time to time to wash away the starch that exudes from it, there will be found to remain a tough, elastic, grayish substance which has been called *gluten*. The presence of at least ten per cent. of this substance in flour is essential to the making of good bread. Besides its gluten and starch, flour contains a small quantity of glucose, which, when mixed with yeast, becomes transformed into alcohol and carbon dioxid. In breadmaking the flour is converted into a stiff dough by kneading with water and yeast; after which it is set aside over night at about 75° Fahr. to ferment. If the temperature be too low, fermentation will not take place; if too high or too long continued the dough will become sour from the transformation of a part of the alcohol into vinegar. During the fermentation the dough rises by the development of carbon dioxid in minute bubbles throughout its substance, each bubble held in position by a cell-wall of the surrounding gluten and starch. The mass is then kneaded over again to perfect the mixture and break up any large cavities which may have formed, and is cut off into loaves of the size desired. These are permitted to stand for half an hour for the continuance of fermentation in the forms, after which they are placed in the oven. The heat expands the minute bubbles of gas which permeate the dough, and the loaf swells correspondingly; but this ceases as the moisture becomes dissipated and the surface hard and brown. During the process the starch-cells swell and burst, and much of their material becomes converted into glucose. Breadmaking transforms a stiff, solid mass of gluten and indigestible starch into a substance so porous that the liquids of the alimentary apparatus can easily penetrate it. It must not be forgotten, however, that all the advantages of good bread are lost if it be eaten while still warm from the oven. It is then in an unfinished state; there is so much moisture and plasticity in it that chewing, instead of breaking it up into minute particles, consolidates it again into a heavy, glutinous mass. By the time it is thoroughly cold this undesirable plasticity is lost. One hundred pounds of flour yield about 133 pounds of bread.

575. Dough may be made porous also by mixing with it certain substances which will permeate its mass with bubbles of carbon dioxid. Carbonate of ammonia is sometimes used for this purpose. The heat of the oven liberates and expands the gas, and at the same time dissipates the volatile ammonia. But most of the *baking-powders*, as these substances are called, consist of cream of tartar and bicarbonate of soda mixed with a proportion of corn-starch to keep them from acting on each other until they are wanted. These add to the bread a certain quantity of Rochelle salt —tartrate of soda and potash—as the result of their combination. Acid phosphate of lime and bicarbonate of soda, with starch as a preservative, are sometimes used, the residual salts in this instance being phosphates of lime and soda. Alum and bicarbonate are also used; part of the sulphuric acid of the alum liberates the carbon dioxid, and alumina remains as an insoluble and probably inert powder along with the sulphates of potash and soda. Leavened or fermented bread is more readily digested than that raised with these powders, as during fermentation starch becomes changed to a considerable extent into dextrin and glucose.

INDEX TO PARAGRAPHS.

A.

Abdomen, compression of, as a mode of artificial respiration, 419
 contusions of, 337
 wounds of, 399
Abscess, 340
Absorbent cotton in burns, 330
 system, 220
Absorption from the stomach, 211
 from the small intestine, 218
 from the large intestine, 219
 by the skin, 237
 of putrefactive products of wounds, 372
 of catgut ligatures, 388, 389
 induced by counter-irritation, 313
Acarus scabiei, 543
Accessory foods, 198
Accidents, etc., management of, 264
 etc., which call for artificial respiration, 420
Accommodation of the eye, 252
Acid phosphate of lime in baking, 575
Acids, corrosive effects of, on the skin, 332
 poisoning by, 500
Aconite as a poison, 520, 507

Aconite as a remedy in poisoning, 530
Acromion process, fracture of, 454, 152
Acting Hospital Stewards, duties of, 9
Actual cautery, in dog bites, 376
Administration building of general hospitals, 114
Administrative system of the human body, 244
Aeration of the blood in the lungs, 225
Æther, inhalation of, 276, 277, 279
Affusion, cold, in asphyxia from noxious gases, 420
Agriculture, a cause of malarious exhalations, 71
Air, carbon dioxid in, 226
 effects of deprivation of, 227
 exposure to, for the arrest of bleeding, 386
 furnace-heated, 113
 respired; effects of breathing it, 228
 respired, upward tendency of, 231
 sanitary analysis of, 229
 volume of, required to ventilate, 230

INDEX TO PARAGRAPHS.

Air cells of lungs, 222
    movement, during ordinary breathing, 225
Air space in log huts for winter use, 93, 96
    in hospital wards, 113
Airing of bedding on transports, 108
    of interior of huts, 94, 99
    ward or barrack-rooms, necessity for, 85
Albumen of the blood, 170
    as a food, 198
    test for, in urine, 297
Albuminoids decomposed into urea, 240
    digestion of, in the small intestine, 217
    of food, solution of, in stomach, 211, 212
    in fresh meat, 199
    in pork, 199
    influence of cooking on, 564
Albuminuria in yellow fevers, 109
Alcohol in serpent bites, 377
    insensibility from, 410
Alimentation, 198
Alkalies, poisoning by, 500
Alum in baking, 575
    as a styptic, 383
Ambulances, cases that require transportation in, 27
Ambulance boxes for emergency supplies during service, 30
    companies, 27
Ambulance officers, duties and responsibilities of, 27, 31, 61

Ambulance stations during battle, 39, 42
    train on the march, 31, 32
American National Red Cross, 56
Ammonia as a counter-irritant, 313, 317, 329
    to the nostrils to promote respiration, 419
    poisoning by, 500
    in serpent bites, 377, 379
Amœboid movements, 172
Amputation, triangular bandage on stump after, 370
Amputations, assistance required during, 273
Anæsthesia, local, 280, 281
Anæsthetics, 276
Anastomosis, 187
    venous, 193, 195
Animal heat, 243
Ankle, sprains of, 338
    dislocation of, 498; 167
Anterior tibial artery controlled by pressure, 367, 182
Antidotes, chemical, in poisoning, 503
Antimony as a poison, 515, 503
Aorta, 178, 181
Apomorphin as an emetic in poisoning, 523, 502, 522
Apoplexy, 403
Arachnoid membrane, 247
Areolar tissue, 143
Arm, fracture of, 455–459, 154
    to arm vaccination, 547
    sling for, 370
Army itch, 544
Arnica in contusions and sprains, 337, 338

Arrow wounds, 373
Arsenic as a poison, 509, 503, 504
Arsenical soap in toothache, 346
Arterial blood, 176, 188, 197
    coats, retraction of, when cut, 385
    hemorrhage, 385
Arteries which can be compressed, 367
    enlargement and contraction of, 185
Artery forceps, 273, 389
Articles of Geneva Convention, 54, 55
Artificial respiration, 416
    respiration, conditions which call for, 420
    respiration in chloroform insensibility, 278
    respiration in cases of poisoning, 507, 508, 522, 523, 526, 527, 531
Asphyxia, 227, 416
    from coal-gas, charcoal fumes, etc., 420
Astigmatism, 257
Astragalus, 168
Astringents for capillary hemorrhage, 383
A-tent, 89
Atropin as a poison, 525
    as a remedy in poisoning, 507, 508, 522, 523, 527
Attending surgeons at field division hospitals, 30
Auditory canal, 259
Auricles of heart, 176
Autumnal fevers, causation of, 64–71

Axilla, 180
Axillary artery, 180
    artery controlled by pressure, 367
    vein, 190

B.

Back-bone, 146
Bacteria of nitrification, 116
    of putrefaction in food, 563
    of putrefaction in wounds, 359, 362, 363
Baked meats, 573
Baking of bread, 574
    powders, 575
Bandage, triangular, 370, 395
Bandages, roller, 439
    starch, etc., 441
    tailed, 441
Base hospitals, 110
Basilic vein, 194
Baths, 270
Bath-tub connections with soil-pipe, 124
Battle, the field hospital during, 35–52
    injuries and loss in, 27
Bayonet wounds, 355
Bed, preparation of, for patient, 268
Bedding of the field hospital, 30, 43
Bed-sores, 293
Bedsteads in temporary camps, 91
Beef essence, 569
    stock, 568
    tea, 569
Bees, stings of, 379

Belladonna as a poison, 525, 508
Bell-traps, 133
Bicarbonate of soda in baking, 575
Biceps muscle, 156, 180, 194
Bicuspids, 206
Bile, 216
Bite of rabid dogs, 376
Bladder, urinary, 239, 161
    hemorrhage from, 295
Blank forms for reports, etc., 14-17
Blankets, airing of, 99
    in the field hospital, 30
Bleeding from wounds, 367, 385
Blisters, 318
    on burned or scalded surfaces, 330
    on the feet, 353
    introduction of medicine by means of, 237
Blood, the, 169-172
    alteration of, in sunstroke, 404
    apoplectic extravasation of, 403
    expectorated, in chest wounds, 397
    flow of, accelerated by muscular action, 195
    in the pleural cavity, 397
    quantity transmitted to an organ, 185, 215
    in urine, 295
    vomiting of, in poisoning, 509, 521
Blood-colored urine in malarial fevers, 295

Blood-corpuscles, renovation of, 215
Bloodless operations, 275
Blood-letting, vein selected for, 194
Boiling as a cooking process, 566
    influence of, on hard waters, 119
    as a mode of purifying water, 106
    in water as a disinfecting measure, 272, 535, 543, 550, 552, 554, 556
Boils, 341
    blind, 342
Bone, 142
    capillaries of, 188
    forceps, 273
Books of record kept at hospitals, 18, 134
Bovine lymph, 547
Bowels, condition of, related to varicose veins, 196
Brachial artery, 180
    artery controlled by pressure, 367
    plexus, 248
Brain, the, 244-247
    compression of, 401
    concussion of, 401
    congestion of, 402
    influence of venous blood on, 227
    exposure of, in wounds, 447
Brassard of hospital corps, 54
Bread-baking, 574
Breast-bone, 151
Brigade hospitals, during war, 24, 25

INDEX TO PARAGRAPHS. 323

Bright's disease, 296
Broiling as a cooking process, 571
Bromid of potassium in poisoning, 508, 531
Bronchial breathing, 225
    tubes, 222
Bruises, 337
Bubo, treatment by pressure, 327
Buffy coat of blood, 170
Bugs in the ear, 429
Bunions, 352
Burns, 329–331

C.

Calcaneum or heel-bone, 168
Camp of field hospital, establishment of, 33, 34
    importance of site of, 62
    change of site needful, 80, 102, 109
Camp-site, selection of, 72, 73, 109
Camp-grounds, old, objectionable, 73
Camp guard of field hospital, 34
Camp diseases, 75–86
Camps, cleanliness of, 101
    sanitary care of, 107
Canine teeth, 205, 206
Cantharides as a counter-irritant, 318
    as a poison, 519, 507
Capelline bandage, 394
Capillaries, 188
Capillary hemorrhage, 188, 382, 383
Carbolic acid in infectious diseases, 536, 537

Carbolic acid as a poison, 522, 504, 507
    acid in burns, 329
    acid in toothache, 346
Carbolized catgut, 388
Carbonate of ammonia in bread-making, 575
    of soda in burns, 329
Carbon dioxid exhaled from the lungs, 197, 229
    insensibility from, 420
    a measure of the vitiation of respired air, 229
    in bread-making, 574, 575
    in air before and after breathing, 226, 229
Carbonic oxid, insensibility from, 420
Carbuncles, 343
Carcasses, disposal of, 103
Cardiac region, 177
Caries, 346
Carotid arteries, 180
    artery controlled by pressure, 367
Carpus, 158
Carron oil in burns, 329
Cartilages of the air tubes, 222
    of the joints, 153
    of the ribs, 151
Casein as a food, 198
    of milk, 200
Cases to be reported, 15
Castor oil in injuries of the eye, 332
Catgut ligatures, absorption of, 388
Cathartics, drastic, as poisons, 517

324   INDEX TO PARAGRAPHS.

Catheter, in brain injuries, 401, 403
    in fracture of the pelvis, 466
    in fractures of the spine, 450
Catlin, 273
Cauterization of poisoned wounds, 375–377
Cellular tissue, 143
Centigrade temperature scale, 287
Centipedes, stings of, 378
Cephalic vein, 194
Cerebral nerves, 244
Cerebrospinal meningitis in overcrowded quarters, 85
Cervical nerves, 248
    vertebræ, 146
Cess-pools, disinfection of, 537
    of general hospitals, 121
Chafing in the groins, 354
Chair for dental operations, 347
Chalk in cases of poisoning, 500, 504, 516
Chaplain of hospitals, duties of, 137
Charcoal in frost-bite, 335
Cheese, poisoning by, 532
Chemical analysis of drinking-water, 118
Chest, triangular bandage for, 370
    wounds of, 397
    wounds, effects of, on respiration, 227
Chicken-pox, 551
Chief Surgeon, duties of, 35, 39, 42, 46
Chilblains, 336

Chill, increased heat of body during, 322
Chloral as a poison, 524, 508
    as a remedy in poisoning, 508, 531
Chlorids in the urine, 241
Chlorinated solutions as disinfectants, 536, 537, 550, 554, 558
Chloroform by inhalation, 276, 278
    in cases of poisoning, 508, 531
    in dislocations, 473, 491
Choking, artificial respiration for, 420
    by fragments of meat, 432
Cholera, 82, 559
Chronic inflammation, 327
Chyle, 218
Chyme, 211, 214, 218
Cicatrix of burns, 330
Cicatrization of wounds, 360
Cineritious matter of the brain, 246
Circulation of the blood, 174
    systemic, 175–196
    pulmonary, 197
    portal, 192
    collateral, 187
Cisterns, 116
Clavicle, 152
    dislocation of, 477
    fracture of, 452, 453
Cleanliness in the kitchen, 563
    in the treatment of wounds, open sores, etc., 282
    personal, in winter quarters, 100
    of troops on transports, 108

Clerical work, 14, 18, 58, 59, 134
Climatic influences, a cause of camp diseases, 75
Clothing Account Book, 14, 18
    fragments of, in wounds, 356
    inadequate, a cause of camp diseases, 75, 76
Clotting of blood, 170, 172
Clove-hitch, 483, 480, 491
Coagulability of fibrin, 170
    of lymph, 220
Coagulated blood, arrest of hemorrhage by, 385
Coagulation of blood, 170, 172
    of chyle, 218
Coaptation of fractures, 436
Cocaine as an anæsthetic, 281
Coccyx, 147
Coffee as a remedy in poisoning, 507, 523, 525
Colchicum as a poison, 518
Cold in the treatment of inflammation, 309
    baths, 270
    climates, food required for, 200
Colic from lead in drinking-water, 120
Collapse, 304
    from loss of blood, 368
    in wounds of chest, 397
    in cases of poisoning, 500, 506, 509, 510, 513, 515, 516–518, 520, 522, 528–532
Collar-bone, 152
    dislocation of, 477
    fracture of, 452, 453
Collateral circulation, 187

Colles's fracture, 460
Colocynth as a poison, 517
Color-blindness, 258
Colorless corpuscles of blood, 172
Comminuted fractures, 434
Common iliac arteries, 181
    tent, 89
Complicated fractures, 434
Complications of wounds of chest, 397
Compound fractures, 434
Compresses, 358
    graduated, 384
Compression of brain, 401, 445, 446
Concave glasses to aid sight, 255
Concentration of troops, effects of, on hospital department, 22
Concussion of brain, 401, 445
Condemnation of medical property, 17
Condiments, 198
Condyles of the femur, 165
    of the humerus, 155, 156
    of the humerus, fracture of, 455, 456
Cone for inhalation of ether, 279
Congestion, 305, 185
    chronic, 327
    of the brain, 402
Congestive fevers, in overcrowded quarters, 85
    malarial fevers, causation of, 64–71
Conjunctivitis, 424, 425
    caused by gonorrhœal matter, 425, 282
Connective tissue, 143

Contagious character of diseases in overcrowded quarters, 85
  diseases, 83–85, 543–560
  diseases should be recognized early, 542
Contractility of arteries, 185
  of muscles, 143–145
Contraction of cicatrix in burns, 330
Contused wounds, 355–363
Contusions, 337
  exudations in, 363
Convex glasses to aid sight, 254, 256
Convolutions of the brain, 245
Convulsions from disordered digestion, 409
  of epilepsy, 407
  of sunstroke, 404
  in teething children, 205, 409
Cooking, 562
Cooks of field hospitals, duties of, 31, 34, 37
Copper, salts of, as poisons, 514
Corium, 233
Cornea, 251, 257
  sloughing of, 425
Corns, 351
Corpuscles of the blood, 171, 172
  of lymph, 220
Corrosive sublimate decomposed by metals, 536
  sublimate as a disinfectant, 536, 537, 550, 552, 554, 555, 558
  sublimate as a poison, 503, 510

Corrosive acids, effects of, on the skin, 332
  action of strong acids and alkalies, 500
Cotton, wadding or batting as a padding for splints, 438
Cotton-wool in burns, 329, 330
Cotyloid cavity, 164
Council on the effects of deceased soldiers, 14
Counter-extension, 467, 473
Counter-irritation in inflammation, 313, 320
  in chronic inflammations, 328
Cowhorn forceps, 347
Cradles, 357
Cranium, bones of, 148
  fractures of, 445
Cream of tartar in baking, 575
Crepitus in fractures, 435
Croton oil as a counter-irritant, 320
  as a poison, 517
Croup, suffocative paroxysms of, 313
Crusta petrosa, 208
Crystalline lens, 251
Cupping-glasses, 312
Cuticle, 233, 234
  development of, in vesicated burns, 330
Cut throat, 396
Cyanid of potassium as a poison, 526

D.

Dead, disinfection of the bodies of the, 536

Dead house of General Hospitals, 114
Deafness attending sore throat, 259
Death from chloroform inhalation, 278
　signs of, 421
　in fractures of the spine, 450
Deaths and interments, Register of, 18
　sudden, in camp, due to overcrowding, 85
　en route to base hospitals, how recorded, 47
Decay of vegetation, influence of, on malarious exhalations, 69
Decomposition as a sign of death, 421
Deficiency of food, effects of, 203
Deformities caused by the contraction of burns, 330
Deformity in fractures and dislocations, 435, 473
Deglutition, 204
Delirium attending hemorrhage, 368
　tremens, 412
Deltoid muscle, 153, 154
Dental pulp, 208, 209
Dentine, 208
Derma, 233
Descriptive Lists and Books, 17, 18
Desquamation after inflammation, 305
　in burns, 329
　in frost-bite, 334

Destruction by fire as a disinfecting agency, 541, 550
Detached troops, hospital provision for, 53
Dextrin, 564, 218
　produced by action of saliva, 210
Diabetes, 296, 298
Diaphragm, 161, 223
Diarrhœa, association of, with scurvy, 77
　causation of, 76, 202
　caused by hard waters, 119
　in teething children, 205
Diet, deficient, effects of, 77, 203
　errors of, a cause of boils, 341, 342
　errors of, a cause of camp diseases, 76
　full, when required, 201
　of invalids, 283
　low, as a remedial measure, 326
　of patients in hospital, 201
　variety of, necessity for, 200
Digestion in the large intestine, 219
　in the small intestine, 217
　in the stomach, 211–213
Digitalis as a poison, 530, 508
　as a remedy in cases of poisoning, 507, 508, 520, 529
Dining-room of General Hospitals, 114
Discharge papers of enlisted men, 14
Discipline of General Hospitals, 136
Disinfection, 272, 535

Disinfection by boiling water, 535, 550, 552, 554, 556
  by carbolic acid, 536, 537
  by chlorinated lime or soda, 536, 537, 549, 554, 558
  by corrosive sublimate, 536, 537, 549, 552, 554, 558
  by destruction by fire, 541, 550
  by dry heat, 539
  by sulphur vapor, 540, 550, 554, 556
  by steam, 539, 550
  in chicken-pox, 552
  in cholera, 559
  of excreta, 536
  in glanders, 381
  of hands, dressings, etc., in surgical work, 272
  in itch, 543
  in measles, 556
  of privy-vaults or cesspools, 537
  in scarlet fever, 554
  in small-pox, 550
  in typhoid fever, 558
  in yellow fever, 559
Dislocations, 472–499
Dislocation, prevention of recurrence, 475–477, 480, 481, 495
  complicated, 474
  symptoms of, 473
  treatment of, 473
Displacement of bone in fractures, 435
  danger of, in fractured pelvis, 466
  danger of, in fractures of the spine, 449, 450

Dissection wounds, 375
Distilled water in cholera or yellow-fever epidemics, 560
Diuretics in inflammation, 321
Division hospital, capacity of, 30
  hospital, personnel of, 30
  hospital, on the march, 31
  hospital in battle, 35–52
Dog bites, 376
Dorsal artery of the foot, controlled by pressure, 367, 182
  nerves, 248
  vertebræ, 146
Double inclined plane for fractures, 469, 471
Douche, cold, in poisoning, 523, 526
Douching for the arrest of bleeding, 386
Drainage, 127
  natural, 63
  for sanitary purposes a cause of malaria, 66, 71
  surface, of camp, 94, 98
Drains of general hospitals, 121
  for sewage and waste water, 122, 127–129
  from wounds, 359
Draughts in ventilation, 230
Dressings, disinfection of, 272
Drill, manual, 19, 20
Drowning, resuscitation of the, 420
Drum-corps, duties of, during battle, 38
Drum of the ear, 259
Dry cupping, 312

INDEX TO PARAGRAPHS. 329

Dryness of camp site essential to health, 63
Dugouts as winter quarters, 95
Dulness on percussion, 161
Dura mater, 247
Dust, influence of, in camps, 73
Dysentery, 81

E.

Ear, 259
    foreign bodies in, 429
    method of syringing, 430
Ears, bleeding from, in fractured skull, 445
Earth-closets, 121
Écraseur, 385
Education of members of Hospital Corps, 12
Effusion of serum, 173
Egg albumen in cases of poisoning, 500, 503, 510, 514, 522
Elbow-joint, dislocation of, 481
    fractures near, 456, 457
Elevators, dental, 349
Emetics, in alcoholic coma, 410
    in opium poisoning, 523
    in poisoning, 502, 504, 509-520, 525-527, 529-532
Emmetropia, 253
Emphysema, 396, 397
Enamel of teeth, 208, 209
Engorgement of blood-vessels, relief of, 307-320
Epidemic disease, specially reported, 15
    disease prevented by cooking, 564
    of measles in newly raised regiments, 84

Epidemic of typhoid fever in newly raised regiments, 80
Epigastrium, 151, 161
Epilepsy, 407, 408
Epistaxis, 283
Epsom salt in poisoning, 504, 511, 521, 522
Equipment of hospital corps, 21
Eruption of chicken-pox, 551
    of measles, 556
    of roseola, 555
    of scarlet fever, 553
    of small-pox, 546
    of the teeth, 205, 206
Erysipelas, 372, 113
Escharotics in poisoned wounds, 395
Esmarch's bandage, 275
Estimates for repairs, etc., to post hospitals, 16
Ether, inhalation of, 276, 277, 279
Eustachian tube, 259
Eversion of the foot in dislocation of the hip-joint, 489
    of the foot in fractured femur, 467
Excreta, disinfection of, 536, 537, 109
Excretion, 221-242
Executive officer of a general hospital, duties of, 134
    officer of field division hospital, 29
Exercise, food required for, 200, 201
    influence of, on the flow of blood through the veins, 195

Exercise, passive, 328
Exhalations, cutaneous, quantity of, 236
  from moist soils, 64, 71
Exhaustion from loss of blood on marches, 415, 32
Experience, importance of profiting by, 23
Expiration, mechanism of, 223
Extemporized general hospitals, 111
Extension in reducing dislocations, 473
  by weights in cases of fractured thigh-bone, 468
Extensor muscles of the upper extremity, 156, 159
  tendon of the thigh, 166
Extensors of the toes, 167
External iliac arteries, 181
  jugular vein, 193
  saphenous vein, 194
Extraction of teeth, 347-349
Extravasation of blood, 173
Exudation, 173
  absorption of, by massage, 328
Eye, 251-258
  drops, mode of using, 423
  foreign bodies in, 422
  injury to, by acids, 332
  teeth, 205, 206
Eyes, glazing of, as a sign of death, 421

F.

Face, bones of, 149
Facial nerve, 249
Fæces, 219
  involuntary passage of, 290, 407, 450
Fahrenheit's temperature scale, 287
Fainting, 302
  on forced marches, etc., 415
Faintness from loss of blood, 266, 368
Farcy, 380, 381
Fat, emulsification of, in the small intestine, 216, 218
  as a food, 198, 199
Fatty layer underlying the skin, 232
Fauces, 204
Felon, 344
Felt, perforated, as a material for splints, 437
Female nurses in general hospitals, 114
Femoral artery, 181, 182
  artery, controlled by pressure, 367
  hernia, 163
  vein, 191
Femur, 165
  fractures of, 467-469
  dislocation of head of, 487-495
Fermentation in the soil a cause of malaria, 65
Festering of wounds, 362
Fever, inflammatory, 322
Fibrin of the blood, 170
Fibrinous principles of food, 198
Fibrous tissue, 144
Fibula, fractures of, 470

Field dressing of gunshot wounds, 369
 hospitals, 22-24
 hospital during battle, 35-52
 hospital on the march, 31-34
 hospital work, object of, 45
 hospitals during winter camps, etc., 57
  Return of killed, wounded, and missing, 59
  Service, organization for, 22-30
 tourniquet, 387
Final statements of enlisted men, 14
Finger nails, growth of, 174
 ligature on, as evidence of death, 421
 pressure for arrest of hemorrhage, 386
 tips, sensitiveness of, 234
 bones of, 158
 dislocation of, 484
 fracture of bones of, 465
Fire, precautions against, in hospitals, 115, 140
First aid in accidents, 264
 aid in gunshot wounds, 365
 dressing-stations during battle, 39, 41, 42
Fish, boiling of, 566
 bones in gullet, 432
Flies, infections carried by, 109
Floating ribs, 151
Flour, use of, in burns, 329, 330
Flour and water in poisoning, 503, 510, 514, 522
 wheat, composition of, 198, 199

Fluctuation, 340
Fly-blisters, 318
Fogs, connection of, with malaria, 67
Foliage, green, influence of, on malarious exhalations, 67, 69
Follicles of small intestine, 214
 of stomach, 211
Food, effects of deficiency of, 203
 effects of excess of, 202
 improper, 76
 of invalids, 283
 propagation of typhoid fever by, 558
 digestibility of different articles of, 212
Foot, dislocations of, 499
 triangular bandage for, 370
Forced respiration, 223
Forearm, fractures of, 460, 463
Foreign bodies, intrusion of, into the system, 422-433
 matter in wounds, 356, 358, 369, 397, 399
Four-tailed bandage, 394, 395, 448
Fractures, 434-471
Fracture-box for fractures of the leg, 471
Fractures, gunshot, of the leg, 471
 progress of union in, 448
 symptoms of, 435
Fresh-air inlet into main drain of buildings, 129
Friction in chronic inflammations, 328
Frost-bites, 334

Fruits, spoiled, bad effects of, 532
Frying as a cooking process, 570
Fund, hospital, 14, 139, 201
Fumigation by burning sulphur, 540, 550, 554, 556

G.

Gall-bladder, 216
Ganglia, sympathetic, 244, 262
Gangrene, in frost-bite, 334, 335
    less common in hand than foot, 187
    hospital, 113
Garters a cause of varicose veins, 196
Gases, asphyxia from noxious, 420
Gastric juice, 211
Gelatin as a nutrient, 568
    in bone, 142
General Hospitals, plans of, 111
Geneva Convention, Articles of, 54, 55
German measles, 557
Germs, destruction of, by cooking, 564
Glanders, 380, 381
Glands, lymphatic, 220
    of the mesentery, 218
    sebaceous, 235
    sudoriparous, 236
Glauber's salt as an antidote in poisoning, 504, 511
Glenoid cavity, 152, 155
Glucose, 564, 574
    produced by action of saliva, etc., 210, 217, 218
Gluteal muscles, 165

Gluten as a food, 198
    of wheat flour, 574, 199
Graduated compresses, 384
Granulation, 359
Granulations, indolent, 361
Gravel, 299
Gravitation, influence of, on the circulation, 308
Gray matter of the brain, 246, 263
    matter of the spinal cord, 248
Grease in kitchen and pantry sink-traps, 124
Green blindness, 258
Gristle, 151
Gumboils, 345
Gums, lancing of, in teething children, 205, 407
Gunshot wounds, 364–371
Gutta-percha filling for teeth, 346, 350
    splints, 437

H.

Hæmoglobin, 171
Hæmorrhoids, 196
Hæmothorax, 397
Hair follicles, 235
Hall's method of artificial respiration, 418
Hamstring muscles, 166
Hand, pronation and supination of, 157, 158
    triangular bandage for, 370
Hands, disinfection of the, 272
Hard corns, 351
Hardness of waters, 119
Head, bandages for, 394, 395
    wounds of, 393
Heart, 175–177

increased action during exercise, 195
cessation of movements, a sign of death, 421
nerves of, 262
Heat as an aid to ventilation, 231
in the preparation of food, 564, 565
in the treatment of inflammation, 310
animal, 243
dry, as a disinfectant, 538
febrile, 322
Heat-stroke, 404
and ventilation of pavilion wards, 112
Hellebore as a poison, 529, 508
Hemiplegia, 291
Hemispheres of the brain, 245
Hemlock as a poison, 527, 508
Hemorrhage, 381-392
arterial, 385
capillary, 188, 382, 383
venous, 384
danger of recurrence during reaction, 390, 303
from collateral circulation, 187, 388, 392
intermediate, 391
nature's modes of arresting, 385
Hemorrhoids, 196
Hernia, 162, 163
Hip, triangular bandage for, 370
Hip-bones, 164
Hip-joint, dislocations of, 485-495
Hopper water-closet, 123
Horses, glandered, 380, 381

Hospital boats and cars, 110
Corps, enlistments for, 3
Corps, instruction of, 4, 12
Corps, organization for field service, 27-30
Corps, reports of, 14
Corps, uniform, etc., of, 21
fever, 85
fund, statement of, 14
gangrene, 372, 113
provision for detached troops, 53
savings, 201
Stewards, appointment and duties of, 7-10
Stewards, duties of, at field division hospitals, 29, 30
Stewards, number of, at division hospitals, 28
Stewards, number of, allowed to posts, 8
Stewards, number of, with ambulances in time of war, 27
Stewards, qualifications of, 10
Stewards' quarters, estimates for, 16
Stewards, re-examination of, prior to re-enlistment, 7
Hospitals, Base, 110
Field, 25-52
General, 110-140
Post, 1-21
Regimental, inefficiency of, 24
Hot baths, 270
water as a counter-irritant, 320

334 INDEX TO PARAGRAPHS.

Houses near field hospitals, 36, 44, 51
Humanized vaccine, 547
Humerus, 155
 head of, dislocation downward, 477–480
 fracture of, 455–459, 154
Huts for troops during winter, 92–96
Hydrochloric acid, burns by, 332
Hyoscyamus as a poison, 525, 508
Hypermetropia, 256
Hypodermatic injection, 502
Hypodermic syringe, disinfection of, 272

I.

Ice, use of, in irritant poisoning, 506
Iced water, objections to use of, in heat-stroke, 406
Identification cards, 15
Ileo-cæcal valve, 219
Ilium, 164
Immovable dressings for fractures, 440–443
 dressings for sprains, 338
Impacted fractures, 434
Incised wounds, 355–357
Incisors, 205, 206
Indolent sores, treatment of, 328
Infection, 543–560
Inferior vena cava, 191, 192
Inflammation, 301–326
 chronic, 327, 328
 local characteristics of, 305
 treatment of, 306–321

Inflammations, special; their symptoms and treatment, 329–354
Inflammatory fever, 322–326
Information slips, 18
Inguinal hernia, 163
Insects in the ear, 429
Insensible perspiration, 236
Insensibility, causes of, 400–415
 by anæsthetics, 278
 from alcohol, 410
 from apoplexy, 403
 from cold, 413
 from concussion of the brain, 401
 from congestion of the brain, 402
 from compression of the brain, 401
 from epilepsy, 407
 from fainting, 415
 from interference with the respiration, 414
 from narcotic poisoning, 523
 from sunstroke, 404
Inspection of troops on transports, 108
 monthly sanitary, 16
 of camps, 107
Inspiration, mechanism of, 223
Instillation of eye-drops, 423
Instruments, surgical, disinfection of, 272
Intelligence, seat of the, 246, 262, 263
Intermediary hemorrhage, 391
Intermittent fevers, causation of, 64, 71
Internal iliac arteries, 181

Internal saphenous veins, 194, 196
Intestines, position of, in abdomen, 161
Inventory of effects, 14
and Inspection Reports, 17
Inversion of the foot in dislocations of the hip-joint, 487, 488
Involuntary muscles, 143
Iodid of potassium in chronic inflammations, 328
Ipecacuanha in poisoning, 502
Iris, 251
Iron, hydrated oxid of, as an antidote in poisoning, 504, 509
Irreducible hernia, 163
Irrigation of wounds, 362
Irritant poisons, 509–518
poisons having a specific action, 519–522
Ischium, 164
Isolation in infectious diseases, 546
Issues, counter-irritation by, 319
Itch, 543

J.

Joints, 153
Jugular veins, 190, 193

K.

Kidneys, position of, 161
hemorrhage from, 295
regulators of the animal heat, 243
Killed, proportion of, to wounded in battle, 27
Kitchen refuse of regimental camps, 103
Kitchen sinks, connection of, with soil-pipes, 124
Kitchens of General Hospitals, 114, 115
Knapsack-room of hospitals, 114
Knee-cap, 166
dislocation of, 496
Knee-joint, 166
dislocation of, 497
sprains of, 338

L.

Lacerated wounds, 358
wounds, relative infrequency of bleeding from, 385
Lacteals, 218, 220
Lactose, 200
Large intestine, 161
Larynx, 150
foreign bodies in, 431
in scalds of throat, 333
suffocation in diseases of the, 227
Latrines, 131
of general hospitals, 121
Laughing-gas, inhalation of, 276
Laundresses, liability of, to typhoid infection, 558
Laundry, connections with soil-pipes, 124
of general hospitals, 114, 115
Lavatories, company, in winter camps, 100
Lead lotions in rubefacience, 329
pipes, action of drinking-water on, 120
subacetate, in eye inflammations, 424

Leaks in plumbing work, 126
Leather splints, 437
Leavened bread, 574
Leeches, 311
Leg, fractures of, 470–472
Lemon juice in cases of poisoning, 500
Lice, 544, 545
Ligaments, 153
Ligation of arteries, 388, 389
Ligatures in poisoned wounds, 375, 377
Limbs, triangular bandage for, 370
Lime-water test for air, 229
Line of battle, first aid given on, 39, 40
Liniments, use of, 328
Lint as padding for splints, 438
Liquor ferri subsulphatis as a styptic, 383
    potassæ in poisoned wounds, 378
    sanguinis, 170
Liston's splint, 467
Liver, actions of, 240
    blood-vessels of, 192
    position of, 161
Location of field hospitals during battle, 35, 36
Log huts as quarters, 93, 96
Loins, sprains of, 339
Long-sightedness, 256
Looking-glass, evidence of life by means of a, 421
Loop bandage, 438
Loss, percentage of, in battle, 27
Lower jaw, dislocation of, 476
Lower jaw, fracture of, 448
Lumbar nerves, 248
    vertebræ, 146
Lungs, capillaries of, 197
    hemorrhage from the, 383
    to remove water from, in cases of drowning, 420
Lymph, 220
Lymphatics, inflammation of, 375
Lymphatic system, 220

M.

Maggots in the nose, 428
Magnesia in cases of poisoning, 500
Malaria, diseases caused by, 64
    generation, evolution, and diffusion of, 64–71
    generated in huts, 95
Malleoli, internal and external, 167
Manipulation in reducing dislocations, 473
Many-tailed bandage, 441
March, field hospital on the, 31–34
Marches in hot climates, 109
Marshall Hall's method of artificial respiration, 418
Massage, 328
Mastication, 204
Mattresses, disinfection of, 541
Measles, 556
    malignant, in overcrowded quarters, 85
    susceptibility of young soldiers to, 84
Meat, cooking of, 566

Meat, dried, 564
> fresh, constitution of, 198, 199
> spoiled, bad effects from, 532

Medical History of Post, 16, 18
> officer the sanitary officer of military camps, 107, 109
> officer, visits of, to the post hospital, 11
> officers, duties of, on the line of battle, 41
> officers, duties of, at ambulance stations, 42
> officers at field division hospitals during battle, 43, 44
> officers in charge of wounded en route to base, 46, 47
> officers supervise the instruction of the members of the Hospital Corps, 12

Medullary matter of the brain, 246

Mercurial vapor-bath, 328

Mercury absorbed by the skin, 237
> administration of, in syphilitic diseases, 238

Mesenteric glands, 218

Mesentery, 214

Metacarpal bones, 158
> bones, fracture of, 464

Metatarsal bones, 168

Meteorological report, 16, 18

Miasms from the human body, 85
> generated in huts, 95

Middle ear, 259

Milk in cases of poisoning, 500, 510, 514, 522

Milk, a perfect food for the child, 200
> teeth, 205

Mineral salts in water supplies, 119

Mists, connection of, with malaria, 67

Moisture needful for the generation of malaria, 65

Molar teeth, 205, 206, 209

Monsel's solution, 383

Monthly Personal Reports, 14
> Reports of Personnel and Transportation, 14
> Reports of Physical Examination of Recruits, 15
> Reports of Sick and Wounded, 15, 59
> Sanitary Reports, 16

Morning Report Book, 18
> Report of the Hospital Corps, 14, 18
> Report of Sick and Wounded, 15, 18

Morphia in cases of poisoning, 500, 506, 509, 515, 517–519, 521, 529

Motor nerves, 249

Mucilaginous drinks in poisoning, 500, 506, 509, 519, 521

Mucous membrane of air passages, 222
> membrane of alimentary canal, 204
> membrane of small intestine, 214, 218
> membrane of stomach, 211
> membrane of the urinary bladder, 239

Mucous membrane, stimulant treatment of, 328
Mucus, 204
Mud in camps, 73
Multicuspids, 206
Muscles, 143
  disintegration of, by use, 240
  spasmodic quivering of, as evidence of life, 420
Muscular coat of arteries, 184, 185, 385
  contraction in cases of fracture, 468
Mushrooms, poisoning by, 532
Musicians during battle, 38
Mustard, counter-irritant action of, 329
  as an emetic in poisoning, 502
  plasters, 316, 313
Muster and Pay Rolls, 14
Myopia, 255

N.

Narcotic poisoning, 523–525, 508
Nasal cavity, maggots in, 428
Nauseants as febrifuge remedies, 321
Navel, 161, 162
Nerves, 244
  of motion, 249
  of sensation, 249
Night air, harmful influence of, 71
Nitrate of silver as a poison, 512, 504
  of silver in poisoned wounds, 375–376

Nitrate of silver to proud flesh, 361
  of silver eye-drops, 424, 425
Nitric acid, burns by, 332
Nitrite of amyl in poisoning, 508, 531
Nitrogenous food, 198
Nitrous oxid gas, inhalation of, 276
Non-nitrogenous food, 198
Nose, bleeding from the, 383
  foreign bodies in, 427
Nurse, value of, depends upon his knowledge, 269
Nurses of field hospitals, duties of, 31, 34, 37

O.

Oakum as padding for splints, 438
Obstruction of veins, 195, 196
Odors, foul, bad effects of, in camps, 76
Œsophagus, 150, 204
Officer of the Day of general hospitals, 136
  of the Day in military camps, 101
Officers' quarters at general hospitals, 114
Oil of bitter almonds as a poison, 526
  olive, in injuries of the eye, 332
  in cases of poisoning, 500, 506, 507, 519, 521
  of vitriol, burns by, 332
Olecranon process, 156
Olfactory nerve, 260

Operating-room in general hospitals, 114
staff of field hospitals, 43–45
Ophthalmia, purulent, 425
Opium as a poison, 523, 508
as a remedy in cases of poisoning, 500, 506, 509, 515, 517–519, 521, 529
Order and Letter Book, 18
Organic matter in drinking-water, 106, 118
matter in the generation of malaria, 65
matter in the perspiration, 236
matter in respired air, 228
Organization for war service, 22
Organs of the body, 141
of the senses, 250–261
Overcrowding, progressive effects of, 85, 86, 97
Overflow-pipes to plumbing fixtures, 124
Oxalic acid as a poison, 516, 504
Oxygen carried by red corpuscles of blood, 171, 188, 197

P.

Pain, cause of, in inflammation, 305
Palmar arterial arch, 180
arches controlled by pressure, 367
Pancreas, 217
Pantry sinks, connection of, with soil-pipes, 124
Pan, water-closet, 123
Papillæ of the skin, 234
of the tongue, 261

Paralysis in apoplexy, 401, 403
of face, 249
in fracture of the spine, 449, 450
in sprains of loins, 339
from sunstroke, 404
Paraplegia, 290
Passive motion in fractures of the leg, 471
Pasteboard splints, 437
Patella, 166
dislocation of, 496
Pathways in camps, 98, 103
Patient, disinfection of, 272
examination of, in accidents, 264
preparation of, for bed, 268, 270
Pavilion wards of general hospitals, 112
Pay accounts of enlisted men, 14
Pectoral muscles, 153, 154
Pediculi, 544, 545
Pelvis, 160
fracture of, 466
Penetrating wounds of the chest, 397, 398
Pensions based upon medical records of war service, 58
Peppermint test for leaky soil-pipes, etc., 126
Peptone, 211, 217
Percussion, 161
Pericardium, 177
Perineal band in fractured femur, 467, 468
Periosteum, 142
Peristaltic movements of the intestines, 216

Peritoneum, 161
Personal reports of medical officers, 13
Perspiration, 236
Persulphate of iron as a styptic, 383
Phalanges of the fingers, 158
    of the toes, 168
Pharmacy, knowledge of, required, 561
Pharynx, foreign bodies in, 432
Phosphates of the urine, 297, 299
Phosphorus, a constituent of nervous tissues, 241
    as a poison, 521, 504, 508
Phrenic nerve, 248, 290
Pia mater, 247
Piles, 196
Pioneers of field hospital, 34, 37, 43
Pit of stomach, 151, 161
Plans of general hospitals, 111
Plantar arch, 182
Plasma, 170
Plaster, application of, to close wounds, 356, 357
    of Paris in fractures, 443, 471
    strapping of sprained joints, 338
    strips in contused wounds, 363
Pleura, 221
Pleurisy, character of respiration in, 227
Plunger water-closets, 123
Pneumonia, condition of air cells in, 227
Pneumothorax, 397
Poisoned wounds, 374, 379

Poisoning by articles of food, 532
    by Rhus toxicodendron, 533
Poisons, classification of, 501
Police, company, 99
    general, of regimental camps, 101, 103
    of general hospitals, 136
Popliteal artery, 182
    artery controlled by pressure, 367
Pork, food principles of, 199
Portal circulation, 192
Posterior tibial artery, 182
    tibial artery controlled by pressure, 367
Post Hospital, service of, 3–21
Potassa as an escharotic, 319
Potassæ, liquor, as a remedy in cases of poisoning, 507
Potatoes, boiling of, 566
Pott's fracture, 498
Poultices, 310
    in abscesses, boils, etc., 340, 343
    after application of a fly-blister, 318
    in corns and bunions, 351, 352
    in frost-bites and chilblains, 335, 336
    in sloughing burns, 330
    in whitlow, 344
    in the treatment of wounds, 361, 375
Presbyopia, 254
Pressure, arteries which may be controlled by, 367
    a cause of inflammation, 352

Pressure, as promotive of absorption, 328
    in the treatment of inflammation, 309
Preventability of camp diseases, 74
Prickly heat, 544
Primary dressing of gunshot wounds, 369
    hemorrhage, 390
    union of wounds, 357
Prison fever, 85
Privy vaults, disinfection of, 537
Processes, spinous, 146
Pronation, 157
    danger of impairment of, in fractures, 463
Property purchased with the hospital fund, 14
    of field division hospital, cared for by, 29
    destroyed, 17, 541
Prostration from loss of blood, 368
Proud flesh, 361
Prussic acid as a poison, 526, 508
Ptyalin, 210
Pubes, 164
Pulleys, use of, in reducing dislocations, 473, 491
Pulmonary arteries, 197, 222
    circulation, 175, 197
    valves, 197
    veins, 222
Pulp of tooth, 208, 209
Pulse, 186, 180
    influence of muscular action on, 324

Pulse, reduction of, by medication, 326
    and respiration, relative frequency of, 223
Punctured wounds, 355, 399
Pupil of the eye, 251
Pupils in conditions of insensibility, 405, 523
Purgatives in inflammation, 321, 326
Purulent conjunctivitis, 425
Pyloric end of stomach, 211, 213

Q.

Quarantine restrictions in cholera and yellow fever, 559, 109
Quarterly report of property, 17
Quiet as a remedial measure, 324
Quinine, use of, 109

R.

Radial artery, 180
    artery controlled by pressure, 367
Radius, 156
    fractures of, 460
Rain conductors, connection of, with main drain, 128, 132
Ration returns, 14
Rations, 201
    for the wounded during and after an engagement, 30
Reaction, general, 304, 305
Record books of a post, 18
Records of field hospitals, 58–60, 30
    medical, importance of, 58

Recruits, report of examination of, 15
Rectum, absorption from, 219
Recurrent bandage for the head, 393
Red-blindness, 258
Red corpuscles of blood, 170, 171, 173
 Cross Society, 56
Redness, cause of, in inflammation, 305
Reducible hernia, 163
Reduction of dislocations, 473
Reef-knot, 389
Reflex action, 262, 263
Refuse matters of camp, disposal of, 103
Regimental hospitals, defects of, in war service, 24
 medical officers, availability of, for special duties, 59
 medical officers, duties of, after a battle, 58
Registers of Patients, 15, 18
 Examination of recruits, 18
 Hospital fund, 18
Remittent malarial fevers, causation of, 64–71
Remittents simulating typhoid fever, 85
Reports and papers from the post hospital, 13–18
 and papers during field operations, 58, 59
 and papers from general hospitals, 134
Requisitions for supplies, 17, 46, 50
Resonance on percussion, 161

Respiration, 223
 artificial, 416–421
Respiratory murmur, 225
Rest in the healing of wounds, 359
 in the treatment of inflammation, 308, 324
Rests during marches, 415
Retching, dry, treatment of, 285
Retina, 251
Retractor, 273
Returns of Medical Property, 17
Revaccination, 549
Reverses in bandaging, 439
Rheumatic inflammations, 327
Rheumatism simulated by scurvy, 77
Rhus, poisonous species of, 533
Ribs, 151
 fracture of, 451
Ridge ventilation, 112
Rigor mortis, 421
Rima glottidis, 224
River water, 106, 117
Roasting as a cooking process, 572
Rochelle salt in bread, 575
Roller bandage, 439
Roseola, 555
Rötheln, 557
Rubefacients, 329
Rupture, 162, 163

S.

Sacral nerves, 248
Sacrum, 147, 160
Saline purgatives, action of, 321
 substances in fresh meat, 199

Saliva, action of, on starch, 210
Salt, Epsom, 504, 511, 521, 523
  Glauber, 504, 511
  table, as an article of food, 198
  table, in poisoning, 504, 512
Salts of the urine, 241, 299
Sand-bags as a support in fracture, 468
Sanitary Report, Monthly, 16
Sanitation of camps, underlying principle of, 74, 85
Saphenous veins, 194
Saw, use of, in amputations, 273
Scabies, 543
Scalds, 330
  of the throat, 333
Scalp wounds, 393
Scalpels, 274
Scapula, 152
  fracture of, 454
Scapular muscles, 153, 154
Scarf-skin, 233
Scarification of the conjunctiva, 425
Scarlatina, 554
Scarlet fever, 553
Sciatic nerve, 248
Sclerotic, 251
Scorpion stings, 378
Screw tourniquet, 387
Scurvy, 77
Sebaceous glands, 235
Secondary hemorrhage, 392
  union of wounds, 359
Sediments in urine, 299
  in water cisterns, 116
Sensations, 244
Sensory nerves, 249

Sequelæ of measles, 84
Serpent bites, 377
Serum, 170
Sewage, 122, 127
  in drinking-water, 106, 118
Sewerage, 127
Sewers of general hospitals, 121
Shampooing in chronic inflammations, 328
Sheets, renewal of, under helpless patients, 292
Shell-fish, poisoning by, 532
Shelter, inadequate, a cause of camp diseases, 75, 76
Shelter-tents, 88
  in field hospitals, 43
Ship fever, 85
Ships, fumigation of, 540
Shock, 302–304
  in burns, 331, 332
  in wounds, 366–368, 397
  death from, 500
Shoes, narrow-toed, bad effects of wearing, 352
Shortening of limb in fracture, 145
Short-sightedness, 255
Shoulder-blade, 152
  fracture of, 454
Shoulder-joint, 153
  dislocations, 478
Sibley tent, 90
Sick and Wounded, Reports of, 11, 15
  and wounded, transportation of, 31, 32, 46
Sieges, field hospitals during, 57
Signs of death, 421

Silicate of soda bandage, 442
Silk ligatures for arteries, 388, 389
Silvester's method of artificial respiration, 417
Simmering as a cooking process, 566, 568, 569
Sinews, 434
Sinks in military camps, 103, 104, 109
Siphonage of traps, 130
Six-tailed bandage for the head, 395
Skin, action of, in regulating heat of body, 243
    desquamation of, in scarlet fever, 329, 554
    inflammation of, in teething children, 205
Skull, bones of, 148
    fracture of, 445–447
    depressed fracture of, 401
Slaughter-house offal in camps, disposal of, 103
Sleep, necessity for, 174, 263
Sleeplessness, caused by hunger, 203
Sling for the upper extremity, 370
Sloughing, in burns, 330
    in frost-bite and chilblains, 334–336
    in contused wounds, 363, 392
Sloughs in boils, 341
Small intestine, 161, 214
Small-pox, 83, 546–550
Smell, the best method of detecting impurity in air, 229
Smith's anterior splint, 469

Smothering, artificial respiration for, 420
Snow-blindness, 426
Snow as remedial in frost-bite, 334
Soap, action of hard water on, 119
    and water in poisoning, 500
Soda, baking or washing, in poisoning, 500
Sod-cloth of tents, 89
Soft corns, 351
Softness of waters, 119
Soil-pipes, 125, 122
    ventilation of, 129
Soils, as relating to the site of camps, 63–73
Sore throat in scarlet fever, 553
Sounds of the heart, 177
Soup, 568
Specific gravity of urine, 296, 298
Spinal cord, 244, 248, 146
    cord, reflex action of, 262, 263
    nerves, 244
Spine, fracture of, 449, 450, 290
    bed-sores in fractures of, 293
Spinous processes of the vertebræ, fractures of, 449
Spirit of camphor in chilblains, 336
Spleen, 215, 161
Splinters in wounds, 356
    a cause of abscess, 340
Splints, 437
    extemporized in field service, 371

Splints, fixed by triangular bandage, 370
   use of, in sprains, 338
   use of, in wounds, 357, 358
Spotted fever in overcrowded quarters, 85
Sprains, 338, 328
Spring forceps, 273, 389
   scarificator, 312
Stable manure, disposal of, 103
Starch as food, 198
   influence of cooking on, 564, 573
   use of, in burns, 329, 330
   bandage, 441, 471
Steam, as a disinfectant, 272, 539
Sterno-mastoid muscle, 150, 193
Sternum, 151
Stewards, duties of, in the field, 29, 30, 37, 41, 47
   at general hospitals, 135
Stewing as a mode of cooking, 567
Stimulant applications in chronic inflammations, 328
   applications in frost-bite and chilblains, 335, 336
Stimulants in shock and syncope, 302, 303, 415
   in prostration from bleeding, 266, 368
   in internal injuries, 337
   in wounds of the chest and abdomen, 398
   in poisoned wounds, 377
   in exposure to extreme cold, 413
   in concussion of the brain, 401
Stimulants in poisoning by alcohol, 411
   in cases of poisoning, 500, 506, 508, 509, 515, 517–522, 526–530, 532
   in artificial respiration, 419
Stomach, 211, 161
   foreign bodies in, 433
   hemorrhage from the, 383
   pump or stomach-tube in poisoning, 502, 509–512, 517, 520, 522–525, 527–529
   nourishment administered by, 396
Store-rooms of general hospitals, 114
Strangulated hernia, 163
Strangulation, artificial respiration for, 420
Strangury in poisoning, 519
Strap-and-buckle fastenings for splints, 438
Strapping with plaster to promote absorption, 328
Straw as a material for splints, 437
Streets, company, proper condition of, 98
Striæ of voluntary muscles, 143
Strychnine as a poison, 531, 508
   as a remedy in cases of poisoning, 508, 524, 528
Styptics, 382, 383
Subclavian arteries, 180
   artery controlled by pressure, 367
Subsistence officer of field hospital, 29
Subsoil water, 106, 117

Sudoriparous glands, 236
Suffocation, 420, 500
Sugar as food, 198
    in urine, 296, 298
    of lead as a poison, 511, 504
Sulci of the brain, 245
Sulphate of copper as an emetic in poisoning, 502, 514
    of copper as an antidote in poisoning, 504, 514, 521
    of copper as a poison, 503, 514
    of copper to proud flesh, 361
    of magnesia as a remedy in poisoning, 504, 511, 521, 522
    of soda as an antidote in poisoning, 504, 511, 522
    of zinc as an emetic in poisoning, 502
    of zinc as a poison, 513
Sulphur, a constituent of albuminous tissues, 241
Sulphuric acid, aromatic, as an antidote in poisoning, 500, 511
Sulphur as a disinfectant, 540
Sunstroke, 243, 404
Superior vena cava, 190
Supination, 157
    in fractures of forearm, 463
Suppression of urine, 300
Suppuration of lymphatic glands, 220
Surgeon in charge of field hospital, duties and responsibilities, 29-60
    in charge of a general hospital, duties and responsibilities of, 134

Surgeon's call at post hospitals, 11
    knot, 389
Surgical operations, preparation for the performance of, 271-279
Susceptibility of young soldiers to typhoid fever, 80
Suspensory bandage for rupture, 163
Suture for closing wounds, 356
Sutures in contused wounds, 363
    not used in wounds of scalp, 393
Sweat, 236
Swelling, cause of, in inflammation, 305
Sympathetic nerves, 244
    nervous system, 262, 263
    fever, 322-326
Symptoms, changes in, noted 265, 269
Syncope, 302, 303, 415
    influence of, on hemorrhage, 385
Synovia, 153
Syphilis, danger of infection from sores, 282
Syphilitic inflammation, 327
Syringing ear, method of, 430
Systemic circulation, 175

T.

Tannin as a styptic, 383
    in cases of poisoning, 503, 515, 518, 525, 527-531
Tarantula bites, 378
Tarsal bones, 168
Tartar on the teeth, 209

INDEX TO PARAGRAPHS. 347

Tartar emetic as a counter-irritant, 320
    emetic as an emetic in poisoning, 502
    emetic as a poison, 503, 515
Tea, green, in cases of poisoning, 503, 515, 518, 525, 527-531
Teeth, 205-209
    extraction of, 347-349
    filling cavities in, 350
Teething, constitutional disturbances attending, 205
Temperature charts, 289
    loss of, as a sign of death, 421
    needful for the development of malaria, 65
    normal, 286
    of patients, how taken, 288
    scales; to change Fahrenheit to Centigrade, and vice versa, 287
Temporal arteries, 179
Tenaculum, 389
Tendo-Achillis, 167
Tendons, 144
    of forearm, 159
Tent-flies for extending hospital shelter, 44
Tents, common, 89
    shelter, 88, 96
    Sibley, 90
Tepid baths, 270
Test for sugar, etc., in urine, 295-300
Tests for organic matter in drinking-water, 362
Testicle, contusions of, 337
Thermometer, clinical, 288

Thigh-bone, 165
Thoracic aorta, 81
    duct, 218
Throat, scalds of, 333
Thumb, bones of, 158
    dislocations of, 483
Tibia, 166, 167
    fractures of, 470, 471
Tibial arteries, 182
    arteries controlled by pressure, 367
Tin, sheet, as a material for splints, 437
Tincture of iodine in chronic inflammations, 320, 328, 329, 336, 352
    of iodine in serpent bites, 377
Tobacco as a poison, 528, 508
Tongue, 261
    appearance of, in disease, 284
    position of, in artificial respiration, 416
Tooth, extraction of, 347-349, 345
    socket, hemorrhage from a, 383
Toothache, 346
Torsion, 386
Touch, special organ of, 234
Tourniquet, 387
Trachea, 150, 222
    foreign bodies in, 431
Transfer slips, 15
Transports, sanitary care of, 108
Traps on fixtures, soil-pipes, and drains, 122-124, 128-130, 133
Traumatic fever, 322-326

Trees, influence of, on malarious exhalations, 67
Trenching of camps, 94, 98, 103
Triangular bandage, 370, 395
Trichiniasis prevented by cooking, 564
Trochanter, 165
Truss, measurement of patient for, 163
Trypsin, 217
Tuberosities of the humerus, 155
Turpentine stupes, 313, 315
  in chilblains, 336
Tympanum, 259
Typhoid fever, infection of, 558
  fever, a camp disease, 79, 109
  fever, germs of, in drinking-water, 106, 558
  fever in overcrowded quarters, 85
  pneumonia due to overcrowding, 85
  seasoning of young soldiers, 80
Typhus fever in overcrowded quarters, 85, 86
  developments in dugouts and badly constructed huts, 95
  prevented by a free allowance of air space, 113

U.

Ulceration of a blood-vessel, 392
  of bunions, 352
  of chilblains, 336
Ulna, 156
  fractures of, 461
Ulnar artery, 180
  artery controlled by pressure, 367
Umbilical cord, 162
Umbilicus, 161, 181
Unguents, use of, in burns, 329, 330
Uniform of Hospital Corps, 21
Union by the first intention, 357
  by the second intention, 359
Units of hospital organization for war service, 22
Uræmic convulsions, 300
Urates in urine, 299
Urea, 242
Ureter, 238
Ureters, hemorrhage from, 295
Urethra, 239
  hemorrhage from, 295
Uric acid, 240
Urinals for winter camps, 105
Urinary sediments, 299
Urine, 240–242, 295–300
  albumin in, 297, 109
  blood-colored, by medicines, etc., 295
  bloody, in cases of poisoning, 519
  bloody, in sprains of loins, 339
  involuntary discharge of, in epilepsy, 407
  retention of, 300
  retention of, in apoplexy, 403
  retention of, in fractures of the spine, 450
  sugar in, 298
  suppression of, 300

Urine, suppression of, in poisoning, 509

V.

Vaccination, 547-549
    of recruits, influence of, on small-pox, 83
Valve water-closet, 123
Valves of aorta, 176, 184
    of heart, 176
    of veins, 195
Vapor-bath, mercurial, 228
Varicose veins, 196
Varioloid, 549
Vegetation, relations of, to malaria, 69
Veins, 189-197
Venæ cavæ, 190-192
Venous blood, 176, 188, 197
    hemorrhage, 384
Ventilation, 230
    of log huts, 93, 94, 96
    of pavilion wards by the ridge, 112
    as a preventive of erysipelas, 372
    of soil-pipes, 129
    of ward in typhoid fever, 558
Vent-pipe in trap to prevent siphonage, 130
Ventricles of the brain, 246
    of the heart, 176
Vertebral column, 146
Vesicants, 314
Vesication in burns, 330
Vesicular breathing, 225
    matter of the brain, 246
Villi of small intestine, 218

Vinegar in cases of poisoning, 500
Vocal chords, 224
Voluntary muscles, 143
Vomiting in alcoholic coma, 411
    in concussion of the brain, 401
    in contusions of the abdomen, 337
    during inhalation of anæsthetics, 277-278

W.

Wagons, army, for transportation of wounded, 46
    field hospital, 30, 31
Walking supervised by reflex action, 262
Wards, fumigation of, by burning sulphur, 540
Wardmasters in general hospitals, 138
Ward Surgeons of general hospitals, duties of, 136
Warm baths, 270
Warm-water dressings, 310
Warming of general hospitals, 112
Warmth obtained at expense of ventilation, 91
    when applied in artificial respiration, 419
Wash-basin, connection with soil-pipe, 124
Washing of clothes in camps, 106
Wasps, stings of, 379
Water, boiled or filtered, to be used, 106, 109

Water, called for in cases of hemorrhage, 368
  cistern, 116
  disinfection of, 106, 109
  dissipation of, in lungs, 197, 226
  douches in narcotic poisoning, 523
  natural purification of, 116
  subsoil, 116
  surface, 117
  tests of purity, 118
  use of, in sunstroke, 405, 406
  use of, in syncope, 415
  well, 117
Water-carriage for sewage, 122
Water-closets, 123
Water-gas, accidents from, 420
Water-seals, 122–124, 128–130
Water-supply of general hospitals, 115
  of camps, 106
  necessity for as a cooling agency, 404
  a cause of camp diseases, 76
  in cholera and yellow fever, 559
  propagation of typhoid fever by, 79, 558
Wedge-tent, 89
Weekly Report of Sick and Wounded, 59
Wells, 117
  in camps, 106
Wet cupping, 312
Whiskey, harmfulness of, in camp, 77
White corpuscles of blood, 172

White matter of the brain, 246
  matter of the spinal cord, 248
Whiting in cases of poisoning, 504, 516
Whitlow, 344
Winds, protection from, in camps, 73
Wire mattress as a surgical bed, 467
Wisdom teeth, eruption of, 206
Wooden splints, 437
Worms, intestinal, prevented by cooking, 564
  convulsions from presence of, in intestine, 409
Wounded, lists of, required after an engagement, 59
  transportation of, to base of supplies, 46
  left to the care of the enemy, 50
Wounds, 355–399
  treatment of, at field dressing-stations, 41, 365
Wrist, fracture of, 460
Wrist-joint, dislocation of, 482

Y.

Yellow fever, 82, 109, 559

Z.

Zinc, action of drinking-water on, 120
  salts of, as poisons, 513
  sheet, as a material for splints, 437
  sulphate of, as an emetic in poisoning, 502

www.ingramcontent.com/pod-product-compliance
Lightning Source LLC
Chambersburg PA
CBHW030253240426
43673CB00040B/957